IET TELECOMMUNICATIONS SERIES 53

Video Compression Systems

Other volumes in this series:

Video Compression Systems

Alois M. Bock

The Institution of Engineering and Technology

Published by The Institution of Engineering and Technology, London, United Kingdom

© 2009 The Institution of Engineering and Technology

First published 2009

The Institution of Engineering and Technology
Michael Faraday House
Six Hills Way, Stevenage
Herts, SG1 2AY, United Kingdom

www.theiet.org

British Library Cataloguing in Publication Data

A catalogue record for this product is available from the British Library

ISBN 978-0-86341-963-8 (paperback)
ISBN 978-1-84919-103-6 (PDF)

Typeset in India by Macmillan Publishing Solutions

To Rosemary and Anita

Contents

Figures

*Figures courtesy of Tandberg Television, part of the Ericsson Group.

Tables

Preface

In the early 1990s, while MPEG was still working towards its first international standard, I gained my first experience of motion-compensated video compression as a development engineer in the Advanced Product Division of National Transcommunications. I was working on a research project to develop a real-time, motion-compensated video compression codec for television transmission. My first impression of such algorithms was that they would be interesting as research projects but that they were far too complex to be used for real-time commercial television applications. Apart from DCT and motion estimation chips, no dedicated ASICs were available to implement the algorithms. Most of the algorithms had to be implemented using earlier versions of FPGAs, comprising a few hundred logic blocks and flip-flops. Nevertheless, in September 1992, even before MPEG-1 had been promoted to International Standard, we had our first real-time SDTV MPEG-1 codec working. It consisted of a 12U encoder and a 3U decoder, and yes, it was a commercial success.

Fast forward to today, one can get an HDTV MPEG-4 (AVC) encoder in a single chip the size of a thumb nail. Video compression has moved not just into television production, broadcast and telecommunications equipment but also into surveillance equipment and a large range of consumer devices: from camcorders, PCs and PVRs to mobile phones.

For quite a few years I have been presenting introductory lectures on MPEG-2 video and audio compression and, more recently, also on HDTV and MPEG-4 (AVC). The courses are organised by the Continuing Education Department of Surrey University and are aimed at engineers with some knowledge of broadcast television. One of the questions I am asked on a regular basis is: 'Are there any books on this subject?' The answer is yes, there are many books on video compression, but most of them are quite mathematical and targeted towards engineers who are developing compression algorithms and equipment. Other books give a brief introduction to digital video processing and compression, but they do not cover the wider issues of video compression systems. This gave me the idea that I could write one myself, aimed at users of compression equipment rather than engineers developing compression encoders.

Acknowledgements

I would like to thank the CEO of Tandberg Television, Part of the Ericsson Group, for giving me permission to publish this book. Furthermore, I would like to express my gratitude to my current and former colleagues at Tandberg Television, NDS, NTL, DMV and IBA who helped me gain experience in video compression technology. Last but not least, I would like to thank my wife for encouraging me to write this book and correcting many typing errors.

Chapter 1
Introduction

Digital video compression has revolutionised the broadcast industry, the way we watch television and indeed the entire infotainment industry, and the changes are still happening: News reports are regularly accompanied by video clips taken using mobile phones. YouTube^{TM} is providing an alternative form of video entertainment. Web 2.0 is changing the way we use the Internet: Users become producers.

One of the major factors of this development was the standardisation of video compression algorithms. This enabled chip manufacturers to invest in video compression technology for consumer applications. It all began in 1984 when the ITU-T Video Coding Expert Group (VCEG) started to develop the H.261 video compression standard. Shortly after, in 1988, the Moving Picture Experts Group (MPEG) was established to develop the first MPEG compression standard. Before that there had been some considerable research into digital video compression but the technology was not advanced enough, and thus the compression efficiency not high enough, for video compression to be used in applications other than high-bit-rate point-to-point video links and studio tape recorders. Once MPEG-1 had reduced movies to bit rates of less than 2 Mbit/s, albeit at a resolution lower than standard television, new applications became possible and MPEG video compression, in its later generations, became a ubiquitous commodity. Today, MPEG and other video compression algorithms are found not just in television studios and telecommunication links but also in camcorders, mobile devices, personal video recorders (PVRs), PCs, etc.

However, to many engineers and managers in the broadcast and telecommunications industry digital video compression has not only provided new solutions but also created many new challenges. Take picture quality as a point in case: Whereas analogue video quality could be fully measured and monitored on a 'live' signal using just four vertical interval test signals (VITS), this is not the case when using digitally compressed video. Although tremendous progress has been made in the last few years, picture quality monitoring and assessment remains a challenge in digital video.

Whereas analogue video signals suffered from noise and interference degradation in the transmission channel, digital video picture quality is largely due to the compression performance of the encoder. Although the data sheet will give some indication of the main features of encoders, it cannot provide enough information to make purchase decisions. The compression operators need to be aware of the impact of important compression tools, such as motion estimation and pre-processing functions, in order to be able to evaluate the performance of encoders under realistic conditions.

The introduction of MPEG-4 advanced video coding (AVC) compression into broadcast and telecommunication systems is adding another level of complexity. Whereas just a few years ago, MPEG-2 was the main video compression algorithm used throughout the broadcast industry, from contribution links right down to direct-to-home (DTH) applications, there are now different algorithms operating in separate parts of the production and transmission path. The high bit-rate efficiency of MPEG-4 (AVC) was the main reason for the rapid deployment of advanced compression systems for DTH applications. However, does that imply that contribution and distribution systems should also change to the new standard or is MPEG-2 the better standard at higher bit rates? And if it is, what are the effects of concatenation of MPEG-2 with MPEG-4?

This book tries to answer these and many other questions. It starts off with a brief introduction to analogue and digital video signals, explaining some fundamentals of video signals, such as the origin and limitations of the interlaced video format, as well as the definition of picture sizes and chrominance formats. These are important properties of video signals referred to in later chapters of the book.

There are, of course, many textbooks available [1–3] explaining the fundamental video and television principles in much more detail, and readers not accustomed with this subject would do well to familiarise themselves with this topic (see Bibliography and Useful websites) before embarking on video compression technology. On the other hand, readers who are quite familiar with video technology and basic compression techniques might want to glance through the first few chapters and concentrate on individual subjects.

Chapter 2 provides a brief introduction to analogue and digital video signals. In particular, it explains the interlaced video format which is highly relevant to video compression algorithms. Furthermore, it outlines video formats used in television and telecommunication systems.

Chapter 3 addresses the problem of picture quality evaluation, an area which is becoming increasingly important in video compression systems. It describes valid applications and limitations of peak signal-to-noise ratio (PSNR) measurements, which are often referred to in this book and in many other publications. Furthermore, it gives a detailed description of objective and subjective video quality measurement techniques.

Chapter 4 gives an introduction to video compression principles. However, it starts off with an introduction to basic audio compression techniques before it describes the fundamental video compression principles in some detail. Audio compression uses techniques similar to those used in video compression, and processes such as transformation, quantisation and masking effects are easier to explain with one-dimensional audio signals than in a two-dimensional video domain.

Building upon the video compression principles explained in Chapter 4, Chapter 5 elaborates on the detailed compression coding tools used in the four currently used MPEG compression algorithms: MPEG-1, MPEG-2, MPEG-4 Part 2 (Visual) and MPEG-4 Part 10 (AVC). Although MPEG-1 is no longer used in broadcast applications, it provides a good introduction to the family of MPEG video standards. Furthermore, MPEG-4 (Visual) is rapidly superseded by MPEG-4 (AVC). Nevertheless, it is useful to provide an overview on all MPEG video standards because it illustrates the development of more and more sophisticated algorithms. In addition to video compression techniques, Chapter 5 also explains the structure of MPEG bit streams and the mechanism for audio–video synchronisation, which is relevant to some of the later chapters.

Having explained MPEG compression standards, one should not ignore other video compression algorithms used in broadcast and telecommunication applications. Chapter 6 describes some of the non-MPEG algorithms available in the public domain. There are, of course, other non-MPEG codecs in use, but many of them are based on proprietary algorithms and it is not always possible to obtain detailed information about them.

One of the most important areas of video compression and video signal processing in general is motion estimation. This is a subject so important that entire books have been written on it [4,5], reason enough to give a brief summary about it in Chapter 7.

Following on the core compression algorithms described in Chapters 4–7, Chapter 8 introduces pre-processing functions, such as picture re-sizing, noise reduction and forward analysis, which are important features of high-end video encoders.

Although the basic video pre-processing and compression algorithms are applicable to all video resolutions, there are important differences when it comes to higher or lower resolutions than standard television. These are explained in Chapters 9 and 10 covering high definition television (HDTV) and mobile applications, respectively.

Having spent more than half of the book on encoding and pre-processing algorithms it is time to have a brief look at decoding and post-processing technologies. Among other decoder features, Chapter 11 explains practical issues such as channel change time and error concealment. This finalises the description of video encoders and decoders. The remaining chapters deal with wider aspects of video compression systems.

Arguably the most important feature of large multi-channel transmission systems is statistical multiplexing. It can provide bit-rate savings of up to 30 per cent virtually independent of the compression algorithm or picture size. After a basic introduction to statistical multiplexing techniques, Chapter 12 explains how such systems can be analysed and evaluated. In particular, it provides a comparison of bit-rate demands of all four combinations of MPEG-2 and MPEG-4 (AVC) at standard and high definition.

Chapter 13 then covers the specific but important application of contribution and distribution systems. Since the requirements for such systems differ quite considerably from those of DTH systems, it is worth investigating the coding tools and encoder configurations for these applications.

One issue of growing importance, already hinted at above, is that of transcoding and concatenated encoding. With the growth of video compression systems, more and more video signals are compressed not just once but many times at different bit rates and/or resolutions. Chapter 14 investigates the effects of concatenation across all combinations of MPEG-2 and MPEG-4 (AVC).

Although decoding and re-encoding is the most general method of re-purposing compressed video signals, there are cases where direct bit stream processing can replace decode–encode combinations. Therefore, Chapter 15 provides a brief description of bit-rate changers and transcoders. Furthermore, Chapter 15 also gives an introduction to MPEG transport stream (TS) splicers, which are often used for the insertion of advertisements.

Although most aspects of this book refer to real-time video compression systems, many of the principles explained are equally applicable to offline encoding systems. Apart from the timing restrictions, the main difference between real-time and offline systems is that the latter can iterate between different modes in order to achieve an optimum result, whereas the former has to get it right first time.

The main sections of each chapter are intentionally written without mathematical equations in order to make them more readable and accessible to a wider audience. Instead, complicated processes, such as motion estimation in the frequency domain, are explained using figures and diagrams with accurate descriptions of the algorithms. However, mathematically minded engineers can still find relevant mathematical explanations at the end of the chapter. For example, the definition of PSNR, referred to above, is explained in Appendix C.

In order to provide some feedback to the reader, there are a number of exercises at the end of some of the chapters. The exercises often refer to more than one chapter, thus highlighting the association between the topics covered in this book. Example answers are provided at the end of the book.

References

1. G.H. Hutson, *Colour Television Theory, PAL System Principles and Receiver Circuitry*, London: McGraw-Hill, ISBN 070942595, 1971.
2. A.L. Todorovic, *Television Technology Demysitfied*, Burlington, MA: Focal Press, ISBN 0240806840, 2006.
3. C. Poynton, *Digital Video and HDTV Algorithms and Interfaces*, San Francisco: Morgan Kaufmann, ISBN 1558607927, 2003.
4. B. Furht, J. Greenberg, R. Westwater, *Motion Estimation Algorithms for Video Compression*, New York: Kluwer Academic Publishers, ISBN 0792397932, 1996.
5. P.M. Kuhn, *Algorithms, Complexity Analysis and VLSI Architectures for MPEG-4 Motion Estimation*, New York: Kluwer Academic Publishers, ISBN 0792385160, 1999.

Chapter 2

Digital video

2.1 Introduction

Digital video is becoming more and more popular. Not only are broadcasters changing over to digital transmission, but most video consumer products, such as camcorders, DVD recorders, etc., are now also using digital video signals. Digitising video signals has many advantages:

- Once digitised, video signals can be stored, copied and faithfully reproduced an infinite number of times without loss of picture quality.
- No further noise or blemishes are added to digital video when it is transmitted from one place to another (provided that there are no bit errors in the transmission).
- Digitally stored video does not deteriorate, even when stored over long periods of time.

But most importantly

- Digital video signals can undergo complex signal processing operations such as standards conversion, re-sizing, noise reduction and, of course, video compression, which cannot be carried out in the analogue domain.

The problem is that digitised video produces vast amounts of data. One second of uncompressed video in standard television format produces 32 Mbytes of data. Therefore, an average movie of 1.5 h would consume 170 Gbytes of disk space. Hence, it is clear that digital video is only feasible in conjunction with video compression. Before we can delve into the world of digital video and compression, we should have a brief look at analogue video signals in order to provide a basic understanding of the origin and limitations of interlaced video.

2.2 Analogue television

When we talk about analogue television, we usually refer to a composite video signal, i.e. a signal in which the chrominance signals are modulated onto a subcarrier. However, on transmission every colour video signal starts off as a three-component analogue signal consisting of the primary colours red, green and blue (RGB). Virtually every colour can be represented as a mixture of these three components. By the time the video signal reaches its final destination at the display, it is once again converted back to these three colours. Yet the RGB signal is not the best way of sending a video signal over long distances because it would require the full bandwidth for all three components.

The first step in converting the RGB signal into a composite signal is its conversion to a YUV signal, where the Y signal corresponds to luma, i.e. brightness, and U and V correspond to two colour-difference signals. Appendix B shows how this conversion is calculated. The reason for this conversion is that although the luma signal requires the full bandwidth of the original RGB signals, the bandwidth of the U and V signals can be reduced without any visual picture degradation, once the signal is converted back to RGB. Having reduced the bandwidth of the colour-difference signals, it is possible to modulate them onto a subcarrier situated towards the high end of the luma bandwidth. The combination of the luma and modulated colour subcarrier signals results in a composite video signal that can be carried in a single coaxial cable requiring no more bandwidth than a single RGB signal. For a more detailed description of composite video signals see Reference 1. Note that the analogue component format YPbPr and its digital version YCbCr are derived from the YUV signal.

There are three basic versions of composite signals: phase alternating line (PAL), National Television Systems Committee (NTSC) and Séquentiel Couleur Avec Mémoire (SECAM). PAL and SECAM are mostly used for 625-line video signals with 25 frames/s [2] and NTSC is used for 525-line video signals with 29.97 frames/s, but there are some exceptions. The subcarrier frequencies in NTSC and PAL, 3.579545 and 4.43361875 MHz, respectively, are carefully chosen to minimise cross interference between the luma and chroma signals.

The fact that both NTSC and PAL signals consist of an odd number of lines per frame is no coincidence. Even the first analogue HDTV format consisted of an odd number of lines per frame (1 125 lines) [3]. This is because halfway through the frame and even halfway through the middle line, the scanning pattern changes from the first field of the frame to the second field. Although this happens during the vertical interval, it can sometimes be seen as half a black line in the active picture area at the top or bottom of a composite signal. The reason for this strange way of changing from one field to the next is to make sure that the lines of the second field are in between the lines of the first field without having to reset either the horizontal scanning pattern or the vertical one. This is the simplest way of producing an interlaced scanning pattern using analogue circuitry. Figure 2.1a gives a graphic representation of a small interlaced frame consisting of 11 lines, and Figure 2.1b shows the corresponding horizontal and vertical scanning waveforms.

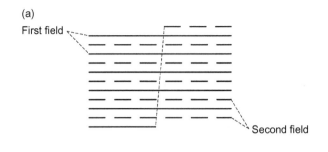

(a)

First field

Second field

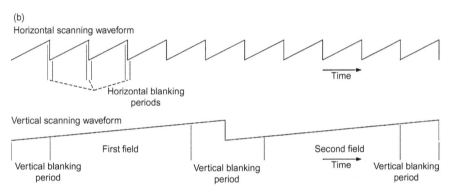

(b)

Horizontal scanning waveform

Time

Horizontal blanking
periods

Vertical scanning waveform

First field

Second field

Vertical blanking
period

Vertical blanking
period

Time

Vertical blanking
period

*Figure 2.1 (a) Scanning pattern of a small interlaced frame consisting of 11
lines. (b) Corresponding scanning waveforms.*

2.3 Interlace

Interlaced scanning reduces the bandwidth requirement of video signals. It
represents a trade-off between vertical and temporal resolution. Slow-moving
material retains the full vertical resolution. As the speed of motion increases,
the vertical resolution of an interlaced video signal decreases. This presents a
good match for cathode ray tube (CRT) cameras and displays because they
tend to produce motion blur on fast-moving objects. With state-of-the-art
solid-state image sensors, the argument for interlaced scanning is not so
obvious because the shutter speed can be fast enough to capture fast motion
without a reduction of resolution. Nevertheless, any reduction in bandwidth
requirements improves the signal-to-noise ratio.

Interlaced video signals, as shown above, are a seemingly very simple
concept. However, if we want to understand the limitations and potential
problems of interlaced signals we need to have a closer look at sampling
theory. Although we are still dealing with an analogue video signal, in order to
understand interlaced video signals properly, we have to take a quick look at
sampling theory because even an analogue video signal is a sampled signal in
the vertical-temporal domain.

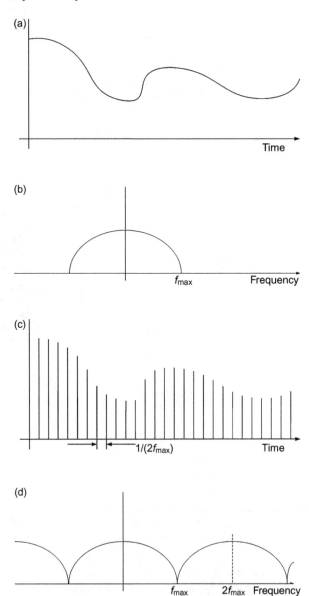

*Figure 2.2   (a) Analogue signal A. (b) Spectrum of analogue signal A.
(c) Sampled signal A. (d) Spectrum of sampled signal A.*

Consider an analogue signal as shown in Figure 2.2a. It might have a frequency spectrum like the one shown in Figure 2.2b, with the highest frequency
being f_{max}. Sampling theory tells us that if we take individual samples of the signal
at a rate of $2f_{max}$ or higher, then we can fully reconstruct the original signal

without any impairments [4]. The reason is that after sampling the signal, as shown in Figure 2.2c, we will get repeat spectra, as shown in Figure 2.2d. It can be seen from Figure 2.2d that if we were to sample at a lower frequency than $2f_{max}$, the spectra would overlap and we would get aliasing artefacts in our signal [5].

Figure 2.3a shows the sampling structure of an interlaced video signal in the vertical-temporal domain. Analogue video lines correspond to individual samples in the vertical-temporal domain. Therefore, the sampling theory explained above also applies to this two-dimensional signal and we get a two-dimensional spatio-temporal spectrum, as shown in Figure 2.3b [6]. We can now see that the highest vertical frequency f_v is identical to the highest temporal frequency f_t of the repeat spectra and vice versa. What does that mean in practice?

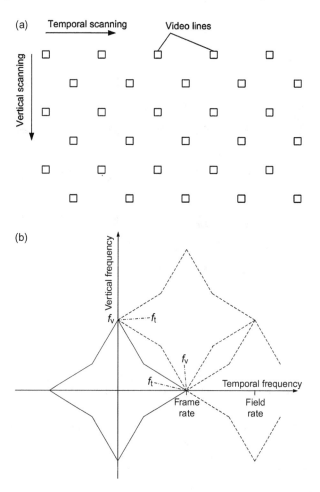

Figure 2.3 (a) Sampling structure of an interlaced video signal. (b) Vertical-temporal spectrum of an interlaced video signal.

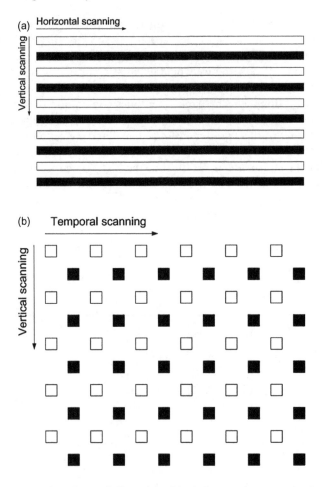

Figure 2.4   (a) White lines, followed by black lines, represent the highest vertical frequency. (b) White fields, followed by black fields, represent the highest temporal frequency.

Consider a video frame that consists of a succession of black and white lines, as shown in Figure 2.4a. This represents the highest vertical frequency in a video frame. Looking at the same signal in the vertical-temporal domain, as shown in Figure 2.4b, we note that the signal consists of a succession of black and white fields, which corresponds to the highest temporal frequency. Therefore, in an interlaced video signal, the highest vertical frequency cannot be distinguished from the highest temporal frequency.

To avoid confusion between the two points, the vertical resolution of interlaced video signals has to be reduced because reducing temporal resolution would lead to unacceptable motion blur. In fact, temporal aliasing is unavoidable in interlaced as well as in progressively scanned or displayed video

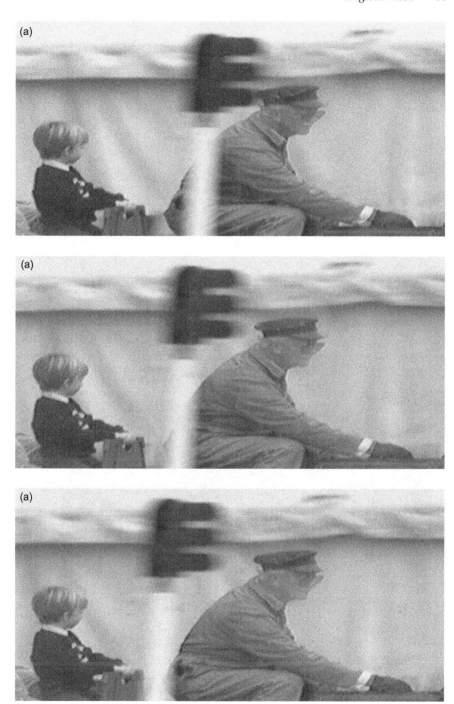

Figure 2.5 (a) Four field pictures. (b) Two frame pictures.

Figure 2.5   Continued

(b)

Figure 2.5   Continued

signals, such as movies. A typical example of temporal aliasing is the so-called waggon wheel effect, where the waggon moves forward but the wheels seem to be turning backwards. Therefore, interlaced video signals cannot utilise the full vertical frequency spectrum, but are limited to about 70 per cent of the theoretical maximum. This is known as the Kell factor [7].

A second, perhaps even more important, consequence of interlaced scanning is that horizontally moving objects generate what looks like vertical detail when viewed as a frame picture. Consider a dark object moving in front of a bright background. As the object moves from field to field, it leaves a vertical frequency pattern at its edges. Figure 2.5a shows a sequence of field pictures of fast pan. Figure 2.5b shows the same sequence as frame pictures. It can be seen that there is a vertical frequency pattern at the edges of the moving objects.

The reason why we went into so much detail about interlaced video spectra is that this seemingly simple method of bandwidth saving causes significant difficulties in standards conversion, picture re-sizing, noise reduction and, of course, video compression [8], as we will see in the later chapters.

Composite video signals using interlaced sampling formats are based on fundamental principles used over the last 50 years in standard definition television (SDTV) signals. However, in the meantime, new video formats have been developed in the computer industry. To understand TV and PC convergence, we need to have a closer look at the various video formats in use today.

2.4   Picture sizes

2.4.1   Standard definition television formats (SDTV)

There are two SDTV formats in use: 625-line 25 frames/s and 525-line 29.97 frames/s. These formats are derived from the analogue composite PAL and NTSC formats, respectively, as explained above. In the digital component format, the two versions are 576i and 480i referring to the number of active video lines and the interlaced format. Both versions are sampled at 27 MHz in 4:2:2 format (see Section 2.5) [9]. This common frequency format formed the basis for the 27 MHz program clock reference (PCR) used in MPEG.

The 625-line format consists of 1 728 samples per line, 720 of which are active luma samples, 288 are horizontal blanking samples and the rest are equally divided between the colour-difference samples Cr and Cb. Vertically, the 625-line format consists of two fields of 288 active lines each; the rest are vertical blanking lines.

Similarly, the 525-line format consists of 1 716 samples per line, also with 720 active luma samples and 360 samples per line for each colour-difference signal. This leaves only 276 horizontal blanking samples. In the vertical direction, the 525-line signal consists of a minimum of 240 active lines per field. If the signal is derived from an NTSC signal, it could have up to 486 active lines per frame but the extra lines are not used in digital compression formats.

Note that neither of the two formats has a pixel-to-line ratio of 4:3, which is the display aspect ratio of conventional SDTV screens. The 625-line format would need an extra 48 luma pixels per line to make square pixels. The 525-line format, on the other hand, has 80 luma pixels, too many per line. This makes these formats unsuitable for a direct 1:1 display on computer monitors because the square pixels in computer monitors would lead to geometric distortions. In practice, this can easily be rectified using sample rate conversion filters (see Chapter 8).

2.4.2   Standard input format (SIF)

The SIF is derived from the SDTV format by down-sampling a SDTV image horizontally and vertically by a factor of two. It was first defined as the nominal, that is standard input format for MPEG-1. Since MPEG operates on 16×16 macroblocks, SIF images are usually reduced to a width of 352 pixels by dropping 4 pixels on either side of the downsized image. Vertically, SIF images consist of 288 and 240 lines for 25 and 29.97 frames/s, respectively.

Although derived from an interlaced SDTV format, SIF and other sub-SDTV formats are usually in progressive format in order to maintain the maximum vertical resolution and because they are often used on PC monitors.

2.4.3   Common intermediate format (CIF)

The CIF format is defined as 288 lines with 352 active luma pixels per line, irrespective of the frame rate. Despite using a common image size at different frame rates, the CIF format, often also referred to as common image format, still consists of non-square pixels and is, therefore, not really suitable for computer displays. In the 25 frames/s domain, SIF and CIF formats are identical; however, at 29.97 frames/s, they are not.

The CIF format was first defined in H.261 for video conferencing applications. In practice, the term SIF is often confused with the term CIF, although nowadays the term CIF is more commonly used. Compression standards usually refer to CIF rather than SIF image sizes because it is a spatially larger image and is identical at all frame rates. If decoders can decode CIF images, then they should also be capable of decoding the smaller SIF images at 29.97 frames/s.

2.4.4   Quarter common intermediate format (QCIF)

A QCIF image is derived by down-sampling a CIF image horizontally and vertically by a factor of 2, resulting in an image size of 176×144 luma pixels. The size of QCIF images is also independent of the frame rate. Note that the QSIF format is rarely used in video compression because in the 29.97 frames/s version it does not divide into an integer number of macroblock rows.

2.4.5   Video graphics array (VGA)

In the previous paragraphs, we have seen that a common sampling rate between different frame rates leads inevitably to different pixel aspect ratios. Even the CIF and QCIF formats, which are independent of the frame rate, do not consist of square pixels. The concept of square pixels was introduced with computer graphics monitors.

Whereas in the television world it was important to maintain the 25 and 29.97 Hz frame rates and find a common sampling rate for digital tape recorders, in the computer world it was more important to use the same image format independent of the frame rate.

The first commonly used video format on computer monitors was the VGA format of 640×480 pixels with an aspect ratio of 4:3. Since then computer monitors have increased their resolutions at an enormous rate, but the VGA format is still a commonly used video format on computer systems.

2.4.6   Quarter video graphics array (QVGA)

In analogy with the QCIF format, the QVGA format is half the width and half the height of a VGA image, that is 320×240 pixels. It is mostly used on mobile devices and for low bit-rate Internet streaming.

2.4.7   High definition television

The first thing to note about HDTV screens is that they are all widescreen with an aspect ratio of 16:9 rather than 4:3. However, in Europe, the 16:9 aspect ratio was already introduced with SD digital television. The real difference between SDTV and HDTV is the higher definition, which produces clearer pictures on large displays. HDTV is also the first television standard that is computer-compatible, since the horizontal-to-vertical pixel ratio is (usually) equal to the geometric aspect ratio of the display device.

There are two versions of HDTV in use on the consumer market today: 1 920 pixels × 1 080 lines using an interlaced scanning pattern, and 1 280 pixels × 720 lines with progressive scanning. The two versions are generally referred to as 1080i and 720p, respectively. Since the progressive format scans every line at every time interval, its frame rate, and consequently its pixel rate, is twice that of an equivalent interlaced format. Therefore, the difference in pixel rate between 1080i and 720p is not as high as it might first appear: 1080i (25 frames/s) has 5 and 720p (50 frames/s) has 4.44 times the pixel rate of a 625-line SDTV signal.

The ultimate HDTV format would be a 1 920 × 1 080 in progressive format with the full 50/59.94 frame rate. This is already being considered for professional production but not for DTH transmissions.

Whereas in SDTV the pixel rate is approximately 10.4 million pixels/s in both video formats (576i and 480i), in HDTV the pixel rate in 29.97 frames/s format is 20 per cent higher than that in the 25 frames/s format. This is because the picture format is the same but the frame rate is different.

In summary, we have seen that in SDTV and its derived formats, the aspect ratio is defined independently of the picture format, whereas video formats developed for PC displays are usually based on square pixels. On the other hand, HDTV formats, which were originally developed for television displays, are equally suitable for PC displays. Note that the RGB-to-YCbCr conversion uses a different matrix in HDTV than in SDTV (see Appendix B). One further differentiation of video formats examined in this chapter is the chrominance format used in video compression systems.

2.5   Chrominance formats

A first step towards reducing the data rate is to convert the RGB signals into a luma signal and two colour-difference signals as mentioned above. Since the spatial resolution of the human eye is lower on colour signals than on luma, the colour-difference signals can be down-sampled by a factor of two without any visual loss of picture quality.

The most common video-component format for professional applications is the 4:2:2 format. In this format, the full vertical resolution is maintained for all components, but the horizontal resolution of the colour-difference signals is reduced by a factor of two. This implies that in SD, the luma signal and the colour-difference signals are sampled at 13.5 and 6.75 MHz, respectively. A 4:2:2 signal is virtually indistinguishable from the full 4:4:4 format, in which all components are sampled at full rate. Figure 2.6a shows the sampling structure of a 4:2:2 video signal.

Originally, the numbers 4 and 2 referred to the absolute sampling rate, i.e. 13.5 and 6.75 MHz, respectively. A progressive video sequence of 720 pixels × 576 lines with 50 pictures/s would have been referred to as an 8:4:4 video format. However, this nomenclature has since been dropped and the numbers 4 and 2 now refer to relative sampling ratios rather than absolute values. For example, an HDTV signal with down-sampled colour-difference signals would also be referred to as a 4:2:2 signal.

In some cases, it is advantageous to down-sample the colour-difference signals even further, for example, by a factor of 3 or even by a factor of 4. These signals would be labelled as 3:1:1 and 4:1:1 formats, respectively [10]. Figure 2.6b shows the sampling structure of a 4:1:1 video signal. Down-sampling the chrominance signals by a factor of 4 results in a noticeable loss of colour resolution. This format is visually no longer indistinguishable from the original 4:4:4 format. However, the advantage is that the full vertical and temporal chroma resolution is maintained. Furthermore, if the video signal had undergone composite coding, the loss in chroma resolution would be minimal. Although 4:1:1 sampling is used in some compression methods, it has not been adopted in MPEG.

A second method of reducing the number of chrominance samples was introduced by the Joint Photographic Experts Group (JPEG) committee. In JPEG, the colour-difference signals are down-sampled both horizontally and vertically [11]. This makes a certain amount of sense because the chrominance resolution of the human eye is similar in horizontal and vertical directions. This format, now referred to as 4:2:0, has the same pixel rate as 4:1:1 but provides a better trade-off between vertical and horizontal chroma resolution. It has been adopted by all MPEG standards. Figure 2.6c shows the sampling structure of a 4:2:0 video signal, assuming progressive scanning.

Although the 4:2:0 format seems to offer the best trade-off between horizontal and vertical resolution, it has its own disadvantages when it comes to interlaced video signals. In principle, there are two methods of down-sampling interlaced chrominance signals:

1. Pretending that the chrominance signal is non-interlaced by dropping the second field and interlacing the first field across both fields. This preserves the maximum vertical resolution but leads to motion judder in highly saturated chrominance areas.
2. Down-sampling the chroma signal field by field to produce an interlaced chrominance signal. This reduces the vertical chrominance resolution but avoids motion judder, and is the preferred method in MPEG. Figure 2.7 shows the vertical-temporal position of the chrominance samples, as recommended by all MPEG standards.

There are, of course, more sophisticated methods of vertically down-sampling interlaced video signals. These are explained in Chapter 8 but are not used for 4:2:0 down-sampling, the reason being that the decoder would

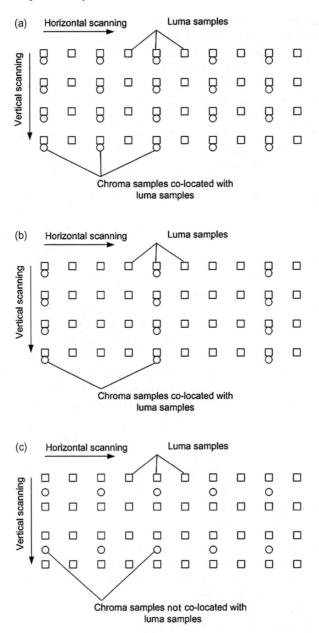

Figure 2.6   (a) 4:2:2 video format. (b) 4:1:1 video format. (c) 4:2:0 video format.

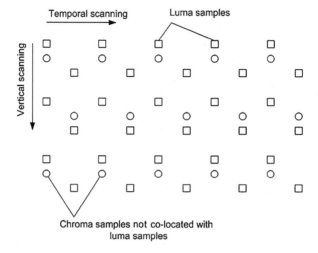

Figure 2.7   Vertical-temporal positions of chrominance samples relative to luma samples in interlaced 4:2:0 video signals

have to implement a matching algorithm to the encoder down-sampling in order to achieve optimum results. Since these algorithms are relatively complex, MPEG did not standardise such methods. Therefore, encoder and decoder manufacturers have not been inclined to implement more sophisticated chroma down-sampling or up-sampling methods.

2.6   Bit depth and uncompressed video bit rates

Although there are some digital composite video formats, they are all but obsolete today. Instead, video signals are digitised in component format, i.e. luma, and the two colour-difference signals are digitised separately and then multiplexed together into a serial digital video signal, commonly referred to as serial digital interface (SDI). The three video components are all digitised with 10 bit accuracy, which is accurate enough to avoid contouring on plain picture areas. In fact, most video compression systems reduce the video signal to 8 bit accuracy before compression. Since a SDTV video signal is sampled at 27 MHz (Y is sampled at 13.5 MHz and Cb/Cr at 6.75 MHz), the corresponding SDI conveys 270 Mbit/s [12]. In HDTV the luma signal and colour-difference signals are sampled at 74.25 and 37.125 MHz, respectively. Therefore, an HD-SDI signal carries 1.485 Gbit/s [13].

2.7   Concluding remarks

In this chapter we have revised some aspects of analogue and uncompressed digital video signals. Although this brief introduction is by no means a comprehensive overview of video technology, it covers some of the topics referred to in later chapters.

2.8 Summary

- In interlaced video signals the highest vertical frequency is identical to the highest temporal frequency. Therefore, the vertical resolution of interlaced signals is not as high as that of an equivalent progressive format.
- Horizontal motion in interlaced video signals appears as high vertical frequencies when viewed as frame pictures.
- All HDTV signals have an aspect ratio of 16:9, but the aspect ratio of digital SDTV can be 4:3, 14:9 or 16:9.
- The most common uncompressed digital video format is 10 bit 4:2:2 in both SDTV and HDTV.
- On the other hand, the most common input format for video compression engines is 8 bit 4:2:0.

References

1. M. Weise, D. Weynand, *How Video Works*, Burlington, MA: Focal Press, ISBN 0240809335, May 2007.
2. PAL-I Standard, 'Specification of Television Standards for 625-Line System I Transmissions in the United Kingdom', DTI Radio Regulatory Division, 1984.
3. J. Whitaker, *Master Handbook of Video Production*, New York: McGraw-Hill, ISBN 0071382461, 2002.
4. R.J. Marks, *Introduction to Shannon Sampling and Interpolation Theory*, New York: Springer, ISBN 0387973913, 1991.
5. B. Mulgrew, P.M. Grant, J. Thompson, *Digital Signal Processing: Concepts and Applications*, New York: Palgrave Macmillan, ISBN 0333963563, September 2002.
6. J. Watkinson, *Television Fundamentals*, Burlington, MA: Focal Press, ISBN 0240514114, April 1996.
7. A.N. Netravali, B.G. Haskell, *Digital Pictures: Representation, Compression and Standards*, New York: Springer, ISBN 030644917X, 1995.
8. J.O. Drewery, 'Interlace and MPEG – Can Motion Compensation Help?', International Broadcasting Convention, Amsterdam, September 1994.
9. K. Jack, *Video Demystified, A Handbook for the Digital Engineer*. 5th edn. Burlington, MA: Focal Press, ISBN 0750683953, June 2007.
10. F.-X. Coudoux, 'Reduction of Color Bleeding for 4:1:1 Compressed Video', *IEEE Transactions on Broadcasting*, vol. 51, Issue 4, December 2005. pp. 538–42.
11. G.K. Wallace, 'The JPEG Still Picture Compression Standard', *IEEE Transactions on Consumer Electronics*, February 1992.
12. SMPTE 259M, '10-Bit 4:2:2 Component and 4fsc Composite Digital Signals – Serial Digital Interface', 1997.
13. SMPTE 292M, 'Bit-Serial Digital Interface for High-Definition Television Systems', 1998.

Chapter 3
Picture quality assessment

3.1 Introduction

Many chapters of this book make references to picture quality. Therefore, it is appropriate to give a brief overview of picture quality measurement methods. For example, statistical multiplexing systems, described in Chapter 12, have to have a measure of picture quality in order to allocate appropriate bit rates to each of the encoders. Although it will be shown in Chapter 4 that the quantisation parameter (QP) is one of the main factors affecting picture quality, there are many other factors influencing the quality of compressed video signals. This chapter gives a brief summary of picture quality assessment methods.

Video compression poses new challenges to picture quality measurements. Whereas in analogue video signals, picture quality could be evaluated with a few, relatively simple measurements, this is not the case with compressed video signals. Measurement of the picture quality of compressed video signals requires extensive knowledge of the human psycho-visual system and the development of complex algorithms in order to achieve a good correlation with the mean opinion score (MOS) of human observers.

3.2 Subjective viewing tests

In the absence of objective measurements, subjective viewing tests can be carried out. These are not only very difficult to set up and require a huge amount of time and effort, but are also not as reliable as one might hope. In order to make the tests as reliable and repeatable as possible, a number of methodologies have been defined [1–3], of which the double stimulus continuous quality scale (DSCQS) method is the most popular.

In this test the observers are shown a pair of sequences A and B, one at a time. Each sequence is 10 s long, followed by a 3 s mid-grey interval. The two sequences are shown twice (ABAB), as illustrated in Figure 3.1, and during the second half of the test, the viewer is expected to mark the picture quality of both sequences within the range 0–100. In order to give the viewer some guidance as to how to judge the picture quality, the range from 0 to 100 is divided into five equally spaced categories, which are labelled as excellent, good, fair, poor and bad, as shown in Figure 3.2.

All viewers should be non-experts, and they should have no prior knowledge of the sequences they are looking at, although a brief introduction of the purpose of the test might be helpful. To calibrate the results, lower and upper anchor sequences are mixed in with the test sequences. For the lower anchor, a worst-case sequence can be used, whereas for the upper anchor, an uncompressed source sequence is usually chosen. These sequences are shown to the viewers in the introduction to give them some indication of the range of picture quality they will be seeing. Furthermore, the upper and lower anchor sequences are used during the test to check the reliability of the viewers. If a viewer marks the upper anchor much lower than a heavily distorted sequence, the results from this viewer should be discarded. The average of all valid results form the MOS of a particular test.

Although the DSCQS test gives reasonably consistent results, it has several disadvantages. To make the tests repeatable, the viewing conditions are exactly defined: the viewing distance from the display (between 4 and 6 times the height of the display), ambient lighting level and so on. These conditions are, however, somewhat artificial and quite different from watching television at home. For a start, there is no audio track on the test sequences, yet it is well

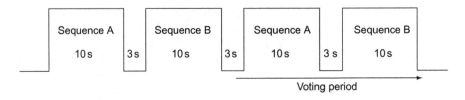

Figure 3.1   DSCQS test

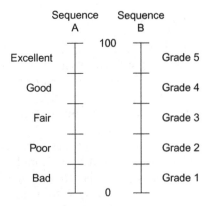

Figure 3.2   Picture quality marking

known that video quality is judged higher if it is accompanied by an audio signal [4]. Furthermore, since the sequences are quite short, there is no storyline to follow, and quite a few sequences are inevitably repeated. Therefore, the viewers get bored quite easily and they might lose attention. It is not unusual for a viewer to lose track of where they are in the test.

For these reasons, a number of other test procedures have been developed. For example, the single stimulus continuous quality evaluation (SSCQE) is a more realistic test [1,5]. More recently, a new test procedure for video quality evaluation has been proposed [6], and research into this subject is ongoing.

3.3 Analogue video quality measurements

Analogue video quality measurements can be divided into three categories: measurements of linear distortions, non-linear distortions and noise. The combination of these measurements provides a complete picture of the quality of analogue video signals.

3.3.1 Linear distortions

Linear distortions are due to irregularities in the frequency response of the transmission channel. Since analogue video signals range from a few Hertz to about 6 MHz, linear distortions are divided into four categories according to the visual distortion effects they cause.

Short-time linear distortions, ranging from a few nanoseconds to about 1 μs, affect the horizontal detail and sharpness of the picture. Linear distortions ranging from 1 to 64 μs cause horizontal patterns. Those ranging from 64 μs to 20 ms cause vertical patterns, and longer ones have temporal effects. Other linear distortion measurements test chroma-to-luma gain inequality and chroma-to-luma delay inequality.

3.3.2 Non-linear distortions

Non-linear distortions are due to irregularities in gain depending on signal levels. The three most important measurements are luma non-linearity, differential gain and differential phase. Luma non-linearity manifests itself as a reduction in contrast towards the white level (white crushing) or towards the black level (black crushing). Differential gain causes changes in colour saturation at different luma levels, whereas differential phase results in hue changes at different luma levels.

3.3.3 Noise

Several types of noise can be distinguished. The three most common types are random Gaussian noise, impulse noise and interference noise. Gaussian noise is caused by thermal noise in the capture device and/or in an analogue transmission path. Impulse noise can be caused by FM threshold 'clicks' or by interference from electrical sparks. Periodic interference noise patterns can be caused by co-channel interference or cross-talk from other signal sources.

There are two types of noise measurement in analogue video signals: weighted and unweighted. Weighted noise measurements take the subjective visibility of the noise into account, whereas unweighted noise measurements make no distinction between different frequencies.

3.3.4   Measurement

To measure these distortions, a number of vertical interval test signals (VITS) have been defined. In 625-line television signals, these test signals are on lines 17, 18, 330 and 331. The test signals consist of staircase waveforms, 2T pulses and frequency gratings to measure both linear and non-linear distortions. Figure 3.3 shows the test signals for 625-line television signals. Since the test signals are transmitted in the vertical blanking interval, analogue video quality measurements can be carried out during live transmissions.

Unfortunately, none of these measurements can be used to measure the picture quality of digitally compressed video signals. For a start, the vertical interval lines are usually not coded in digital video signals. Even if they are, such as in an extended 422P@ML MPEG-2 bit stream, they would not provide any information on the picture quality of the main video signal because static test signals are readily compressed with little or no distortion, whereas complex moving images can suffer impairments if the bit rate is too low. In other words, distortion measurements on static test signals would not be correlated to the picture quality of moving images. To measure the picture quality of compressed video signals, an entirely different approach is required.

3.4   Digital video quality

3.4.1   Peak signal-to-noise ratio (PSNR)

Once digital video signals are compressed, their quality is more difficult to evaluate. The simplest, but also the potentially most misleading method is the PSNR measurement. For measuring PSNR, first the mean square error (MSE) of the compressed video signal (after decoding) is calculated as compared to the source signal. PSNR is then calculated as the ratio between the peak signal (255 in an 8 bit video signal) and the square root of the MSE (see Appendix C). It is usually expressed in decibel.

The advantages of PSNR measurements are that they are easy to carry out, sensitive to small changes in picture degradation and accurately reproducible. This makes PSNR measurements very popular when publishing simulation and measurement results. The disadvantage is that the actual PSNR number bears little relationship to the amount of visible distortion [7]. This is illustrated in Figure 3.4, where the rate-distortion curves of two sequences encoded with two types of encoders are shown.

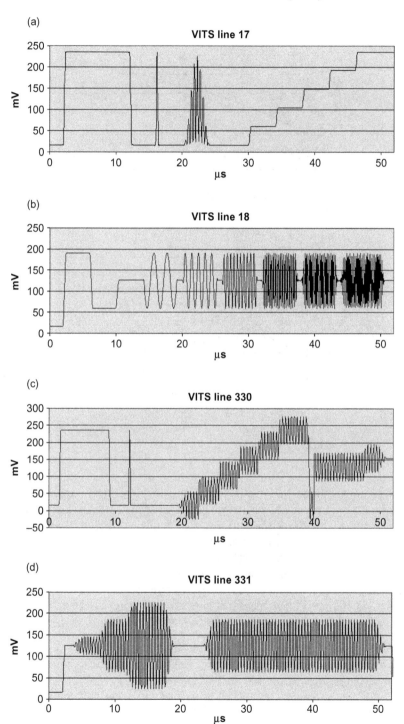

Figure 3.3 Vertical interval test signals (VITS) for 625-line television signals (attenuated)

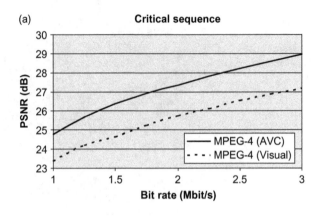

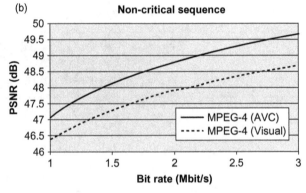

Figure 3.4   Codec comparison of (a) a critical sequence and (b) a non-critical sequence

Figure 3.4a shows the PSNR curves of a highly critical sequence encoded with two types of encoders: a state-of-the-art real-time MPEG-4 (AVC) encoder and a software MPEG-4 (Visual) encoder. Similarly, Figure 3.4b shows the corresponding curves of a non-critical sequence. It can be seen that the PSNR numbers are vastly different and bear no relationship to the visual picture quality. On the other hand, it is possible to obtain a reliable estimate of the bit-rate saving of the MPEG-4 (AVC) encoder compared to the MPEG-4 (Visual) encoder. Therefore, PSNR measurements are advantageous when making picture quality comparisons, provided the following criteria are observed.

- The DC levels of the source signal and the compressed video signal have to be the same. Even a slight difference in DC levels reduces the PSNR value, but has no effect on visible distortion.
- PSNR measurements should be used for comparing only similar types of distortions. A PSNR comparison between two different motion estimation algorithms within the same compression algorithm can provide valuable

information, whereas a PSNR comparison between a digital and an analogue video transmission system, for example, is meaningless.

- The distribution of the error signal should be considered. Two test sequences with similar error signal distributions can be compared using PSNR measurements. If, on the other hand, one test sequence has a small, concentrated area of high distortion and another has a widely distributed area of low distortion, then a PSNR measurement provides no information on the relative picture quality of the two sequences.

- In general, the higher the picture quality and the lower the error signal, the more reliable the PSNR measurements. Badly distorted images should not be evaluated using PSNR measurements. For example, PSNR measurements cannot be used to decide whether blocking artefacts or loss of details is more acceptable.

- Last but not least, there is a relationship between PSNR reliability and picture size. On small images, for example CIF or even smaller images, the correlation between PSNR and subjective picture quality is generally higher than on large images, for example, HDTV-sized images. This is because large images are generally watched with a wider viewing angle. Therefore, the observer will concentrate on certain areas of attention and judge the picture based on these areas rather than the entire picture.

3.4.2 Objective perceptual video quality measurement techniques

We have seen in Section 3.4.1 that simple PSNR measurements can provide useful information in specific cases, as long as they are used under carefully controlled conditions. Nevertheless, PSNR measurements have severe limitations with regard to perceptual video quality measurements. For this reason, a vast amount of research has been carried out, in order to develop more realistic video quality measurement algorithms [8–11].

The Video Quality Experts Group (VQEG) [12] distinguishes between three types of algorithms with different degrees of measurement accuracy when compared to subjective quality tests.

- *Single-ended measurements*: These are used in cases where no reference video signal is available. In this case, the measurement device has to estimate video quality without knowing what the original source signal was like.

- *Double-ended measurements*: These make use of a full reference signal. By comparing the distorted signal with the source signal, double-ended measurements achieve a higher accuracy than single-ended measurements. More importantly, double-ended measurements produce reliable and repeatable results.

- *Double-ended measurements with reduced reference*: These have access to some information about the source, but a full reference video signal is not available. The accuracy of double-ended measurements with reduced reference is generally higher than that of single-ended measurements, but not as high as double-ended measurements.

The three types of algorithms have different applications. Single-ended measurements can be used at the receiving end of distribution systems, in order to monitor video quality in real time. Double-ended measurements are used to make more accurate measurements under laboratory conditions. Double-ended measurements with reduced reference are used in specific cases where some information about the source can be provided, but a full reference signal is not available.

Double-ended measurements can achieve the highest degree of accuracy because they have full access to the reference signal. Although there are many different algorithms, the principle of double-ended measurements is as follows: After calculating a raw error signal as the absolute difference between the distorted and the reference signal, the error signal is filtered in the spatial as well as the temporal domain. This mimics the spatial and temporal masking effects. The difficulty is to filter the error signal in such a way that the total energy of the filtered error signal is proportional to the visibility of the difference between the reference and the distorted signal. In other words, the filtered error signal should be proportional to the visible distortion of the compressed video signal, assuming that the reference signal itself is not distorted.

The results are presented either as relative or as absolute numbers. Examples of the former are just noticeable difference (JND), picture quality rating (PQR) and difference in mean opinion score (DMOS). Absolute picture quality is usually expressed as an MOS between 0 and 100, related to subjective measurements, as explained in Section 3.2 above. JND is a general psychophysical measurement unit defined as: 'The difference between two stimuli that (under properly controlled experimental conditions) is detected as often as it is undetected', whereas PQR is a proprietary measurement unit [13]. DMOS is simply the difference between two MOS values.

Despite significant progress in picture quality analysis over the last few years, the correlation between objective perceptual video quality measurements and subjective viewing tests is not always satisfactory. However, this might be partially because subjective viewing tests themselves are not always very reliable.

3.4.3   Picture quality monitoring using watermark signals

Watermark signals are low-level noise signals superimposed onto the video signal in such a way that it is not visible to the human observer but can be detected by a watermark analyser. Watermarking is used in digital video signals for copyright protection. More recently, there have been proposals to use semi-fragile watermark signals for perceptual picture quality evaluation [14–16].

The watermark signal is masked by the spatial activity of the video signal. The higher the spatial activity, the easier it is to insert, hide and subsequently recover the watermark. High spatial activity, however, masks not only the watermark, but also compression artefacts. In other words, video compression affects the watermark and the picture quality in similar ways. Therefore, the watermark can be used as a reference signal to measure the effect of video compression. By using semi-fragile watermarks, it is possible to use the residual watermark at the receiver as an indicator of perceptual picture quality. This represents a promising approach to single-ended on-line monitoring of digital video transmission systems.

3.4.4 Testing picture quality analysers

With the proliferation of picture quality analysers and algorithms, it is important to test the analysers before relying on their results. A proper comparison would require a full subjective DSCQS test as defined in Reference 1 in order to obtain the MOSs of all test sequences. Subjective viewing tests are not only very time consuming but also very expensive to set up. Alternatively, simpler tests can be carried out, as outlined in the following paragraphs.

A basic test is to encode a test sequence at different bit rates and verify that the analyser reports higher picture qualities for higher bit rates. This test can be repeated for several test sequences, and we should find that different test sequences produce different picture quality results at the same bit rate.

To compare the picture quality of different test sequences encoded at the same bit rate, however, is much more difficult. One possible solution is to carry out a relatively simple test, as explained next.

Take a small number of test sequences encoded at the desired bit rate. Make sure that the test sequences contain different amounts of spatial activity (plain areas and areas with high detail) and temporal activity (fast and slow motions). Then compare each sequence against all other sequences and decide in each case which of the two sequences is less distorted. Note that the number of comparisons to be carried out is $(N \times N - N)/2$, where N is the number of test sequences. In most cases, five test sequences should be sufficient to provide an indication of the reliability of the picture quality analyser. This would require ten side-by-side comparisons of the test sequences.

With each comparison, the test sequences accumulate points. The sequence with higher picture quality gets two points, whereas the sequence with more compression artefacts gets zero point. If two sequences have similar amounts of distortion, they both get one point. This has the advantage that the viewer simply decides which sequence he/she prefers rather than having to decide on an absolute quality rating for each test sequence. The tests should be carried out by several viewers. After adding up all the points for each sequence, the sequences can be put in rank order and the results of these subjective tests can be compared with the test results of the picture quality analyser. If the results correlate, the analyser is trustworthy.

The main advantage of this test procedure is that in each case only two sequences are compared with each other. This is much easier to carry out than trying to allocate an absolute quality number to each sequence. Therefore, the test can be performed by expert as well as non-expert viewers.

An even simpler approach is to use the picture quality results of the analyser as a starting point. By putting the sequences in rank order, one can compare pairs of sequences to see if the visual quality agrees with the results of the analyser.

There are, of course, many different methodologies to test picture quality analysers. However, from the two examples discussed in the previous paragraphs, it is clear that the reliability of picture quality analysers can be evaluated without having to perform a full subjective DSCQS test.

3.4.5   Video quality measurements inside encoders or decoders

3.4.5.1   Double-ended video quality measurements in encoders

The best place to carry out picture quality measurements of a compressed video signal is actually inside a compression encoder. Not only are both the reference and the reconstructed video signals available, but a quality algorithm inside an encoder also has access to the QP, as well as information on spatial and temporal activities, i.e. motion vectors [17]. Quantisation and spatial and temporal activities, however, are the most important factors affecting video quality, since distortion is caused by quantisation and masked by spatio-temporal activity. Some high-end broadcast encoders in statistical multiplexing systems are making use of these factors, in order to calculate the picture quality reported to the multiplex controller (see Chapter 12). After receiving picture qualities from the encoders, the statistical multiplex controller can make sure that each channel is allocated the required bit rate [18].

3.4.5.2   Single-ended video quality measurements
on compressed bit streams

A single-ended version of the algorithm can also be implemented using a decoder. Apart from the video source, a decoder has access to the same parameters as an encoder, i.e. quantisation, spatial activity (derived from the decoded image) and temporal activity (derived from the motion vectors). Unfortunately, in heavily compressed video sequences, the reconstructed picture deviates considerably from the source. Consequently, it is more difficult to estimate spatial activity and the resulting masking affect in a decoder than in an encoder. Nevertheless, single-ended picture quality and PSNR estimates [19] with reasonable accuracy can be carried out on compressed bit streams.

3.5   Concluding remarks

In this chapter we have seen that digital video compression presents significant challenges in terms of measuring and monitoring the quality of service. Whereas in analogue video systems, picture quality could be measured on live video signals at each point in the system using VITS, this is not possible in digital systems. Single-ended picture quality monitors can provide at least a partial solution, insofar as they are able to detect badly distorted signals. A promising solution might be to use watermark signals for picture quality monitoring.

As far as laboratory measurements are concerned, there are three possibilities. Although there are many pitfalls, we have seen that simple PSNR measurements can be used to make picture quality comparisons under carefully controlled conditions. In cases where PSNR measurements are not applicable, we can either resort to subjective viewing tests or use one of the state-of-the-art picture quality analysers that are now available.

There has been significant progress in the development of double-ended picture quality algorithms. State-of-the-art picture quality analysers are now sophisticated enough so that the results could be almost as trustworthy as mean opinion scores of DSCQS tests. However, before a picture quality analyser is used to evaluate compression systems, it is advisable to test the reliability of the analyser. For this reason, a few examples of how to test picture quality analysers, without having to carry out full DSCQS subjective viewing tests, have been explained.

3.6 Summary

- Analogue video quality measurements are not suitable for measuring the picture quality of digitally compressed video signals.
- PSNR measurements can be used under carefully controlled conditions to compare video distortions of a similar nature.
- Picture quality measurements of double-ended picture quality analysers are becoming more reliable.
- Picture quality analysers within encoders have access to the source (spatial and temporal activities), the reconstructed image (distortion) as well as the QP in order to calculate visual quality.

References

1. ITU-R BT.500-11, 'Methodology for the Subjective Assessment of the Quality of Television Pictures', 2002.
2. T. Yamazaki, 'Subjective Video Assessment for Adaptive Quality-of-Service Control', IEEE International Conference on Multimedia and Expo, August 2001. pp. 397–8.
3. I.E.G. Richardson, C.S. Kannangara, 'Fast Subjective Video Quality Measurement with User Feedback', *Electronics Letters*, Vol. 40, Issue 13, June 2004. pp. 799–801.
4. M.R. Frater, J.F. Arnold, A. Vahedian, 'Impact of Audio on Subjective Assessment of Video Quality in Video Conferencing Applications', *IEEE Transactions on Circuits and Systems for Video Technology*, Vol. 11, Issue 9, September 2001. pp. 1059–62.
5. P.N. Gardiner, M. Ghanbari, D.E. Pearson, K.T. Tan, 'Development of a Perceptual Distortion Meter for Digital Video', International Broadcasting Convention, Amsterdam, September 1997. pp. 493–7.
6. H. Hoffmann, T. Itagaki, D. Wood, 'Quest for Finding the Right HD Format, A New Psychophysical Method for Subjective HDTV Assessment', International Broadcasting Convention, Amsterdam, September 2007. pp. 291–305.

7. E.-P. Ong, W. Lin, Z. Lu, S. Yao, M. Etoh, 'Visual Distortion Assessment with Emphasis on Spatial Transitional Regions', *IEEE Transactions on Circuits and Systems for Video Technology*, Vol. 14, Issue 4, April 2004. pp. 559–66.
8. Z. Wang, A.C. Bovik, H.R. Sheikh, E.P. Simoncelli, 'Image Quality Assessment: From Error Visibility to Structural Similarity', *IEEE Transactions on Image Processing*, Vol. 13, Issue 4, April 2004. pp. 600–12.
9. T. Hamada, S. Miyaji, S. Matsumoto, 'Picture Quality Assessment System by Three-Layered Bottom-Up Noise Weighting Considering Human Visual Perception', *Proceedings of SMPTE*, 1997. pp. 20–26.
10. L. Lu, Z. Wang, A.C. Bovik, J. Kouloheris, 'Full-Reference Video Quality Assessment Considering Structural Distortion and No-Reference Quality Evaluation of MPEG Video', IEEE International Conference on Multimedia and Expo, August 2002. pp. 61–64.
11. Z. Wang, L. Lu, A.C. Bovik, 'Video Quality Assessment Based on Structural Distortion Measurement', International Conference on Image Processing, 2002. pp. 121–132.
12. ITU-T J.144 'Objective Perceptual Video Quality Measurement Techniques for Digital Cable Television in the Presence of a Full Reference', 2001.
13. K. Ferguson, 'An Adaptable Human Vision Model for Subjective Video Quality Rating Prediction Among CIF, SD, HD and E-Cinema', International Broadcasting Convention, Amsterdam, September 2007. pp. 306–13.
14. P. Campisi, G. Giunta, A. Neri, 'Object-Based Quality of Service Assessment Using Semi-Fragile Tracing Watermarking in MPEG-4 Video Cellular Services', IEEE International Conference on Image Processing, Vol. 2, pp. 881–4, Rochester, 22–25 September 2002. pp. II/881–4.
15. P. Campisi, M. Carli, G. Giunta, A. Neri, 'Blind Quality Assessment System for Multimedia Communication Using Tracing Watermarking', *IEEE Transactions on Signal Processing*, Vol. 51, Issue 4, April 2003. pp. 996–1002.
16. S. Bossi, F. Mapelli, R. Lancini, 'Semi-Fragile Watermarking for Video Quality Evaluation in Broadcast Scenario', IEEE International Conference on Image Processing, September 2005. pp. I/161–4.
17. L. Boch, S. Fragola, R. Lancini, P. Sunna, M. Visca, 'Motion Detection on Video Compressed Sequences as a Tool to Correlate Objective Measure and Subjective Score', International Conference on Digital Signal Processing, July 1997. pp. 1119–22.
18. A.M. Bock, 'Factors Affecting the Coding Performance of MPEG-2 Video Encoders', International Broadcasting Convention, Amsterdam, September 1998. pp. 397–402.
19. A. Ichigaya, M. Kurozumi, N. Hara, Y. Nishida, E. Nakasu, 'A Method of Estimating Coding PSNR Using Quantized DCT Coefficients', *IEEE Transactions on Circuits and Systems for Video Technology*, Vol. 16, Issue 2, February 2006. pp. 251–9.

Chapter 4

Compression principles

4.1 Introduction

Chapter 2 introduced some important aspects of analogue and digital video, particularly relevant to video compression. In this chapter we have a first look at some basic video compression techniques before we move on to specific MPEG algorithms. Although some of the examples are already based on MPEG-2 coding tools, the principles explained in this chapter are applicable to the majority of video compression algorithms in use today. But before we explore the vast area of video compression methods, we will have a brief look at audio compression techniques.

4.2 Basic audio compression techniques

It may seem strange to start a discussion on video compression with some basic principles of audio compression techniques. There are, however, a number of reasons why this might be a good approach: historically, audio compression was developed before video compression, and some of the principles of audio compression have been 'translated' directly from one-dimensional audio signals to two-dimensional video compression. Furthermore, since an audio signal is one-dimensional, some of the techniques are easier to explain in terms of audio compression than in terms of video compression. Last but not least, it is useful to understand some of the basic principles of audio compression because most video encoders also use audio compression.

In order to develop an audio compression system, we need, first of all, to understand the characteristics of the human psycho-acoustic system. The three most important factors to investigate are the frequency response of the human ear, the masking effect in the frequency domain and the masking effect in the time domain; the masking effect in the frequency domain is probably the most significant factor that audio compression systems can make use of.

4.2.1 Frequency response

The psycho-acoustic principle is based on studies of the audibility of sounds in the presence of other sounds. Figure 4.1 shows the frequency response of the human ear. This is sometimes referred to as the 'Fletcher–Munsen' perception curve [1], named

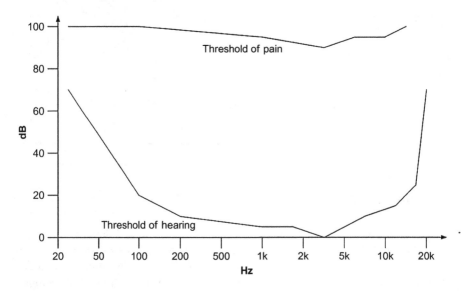

Figure 4.1   Frequency response of the human ear

after the researchers who first measured the sensitivity of the human ear in 1933. The curves indicate levels of equal perceived loudness at different frequencies. The figure illustrates the two extremes: the quietest sounds perceivable, that is the threshold of hearing, and the loudest sounds bearable, that is the threshold of pain.

It can be seen in Figure 4.1 that the quietest sound perceivable is at around 3 kHz, the region that is most important for understanding human speech. Above and below 3 kHz, the human ear is less sensitive. This variation in sensitivity of the human ear can be exploited by using coarser quantisation levels in those areas where the ear is less perceptive. The audio spectrum can be divided into a number of frequency bands, which can be individually quantised. Those bands where the ear is less sensitive can be coded with fewer bits, without having a perceptible impact on audio quality.

To cover a frequency range of 20 kHz, a minimum sampling rate of 40 kHz is required, as explained in Chapter 2. In practice, music CDs use a sampling rate of 44.1 kHz, whereas most broadcast systems use a sampling rate of 48 kHz.

Figure 4.1 also illustrates that the difference between the quietest sound perceivable and the loudest sound bearable is about 100 dB. To cover this vast dynamic range, a minimum accuracy of 16 bits is required. Most state-of-the-art audio compression systems, however, now have an accuracy of 20–24 bits. By comparison, a video signal requires a minimum dynamic range with an accuracy of only 8 bits.

4.2.2   Frequency masking

The shape of the curve representing hearing sensitivity changes in the presence of other sounds. If there is a tone with a certain amplitude at a particular frequency, then hearing sensitivity is reduced in the region around this tone [2]. This is

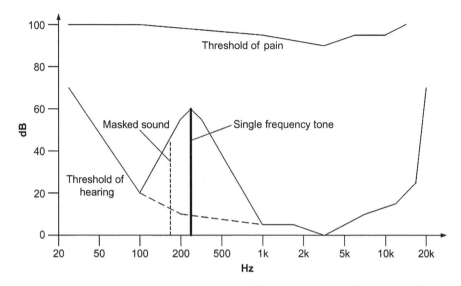

Figure 4.2   Frequency masking effect

illustrated in Figure 4.2, which shows a tone at 250 Hz. Due to the presence of this tone, hearing sensitivity is reduced between 100 Hz and 1 kHz. Any sound in the frequency range near the tone must now be louder than the new curve in order to be audible. Sounds that are below the curve are masked by the presence of the tone and are therefore not audible.

This frequency masking effect is used in audio compression by varying the quantisation in each frequency band according to the residual sensitivity. Figure 4.3 shows how the masking threshold is raised above the noise level due to the presence of masking tones in adjacent frequency bands. Therefore, the two frequency bands shown in this figure can be quantised down to the signal-to-masking ratio, rather than the signal-to-noise ratio that would be required if the two tones were not present.

4.2.3   Temporal masking

In addition to the frequency masking effect, there is also a temporal masking effect [3]. Figure 4.4 shows that about 5 ms before and up to 100 ms after a loud tone, quieter tones below the masking curve are not audible. The temporal masking effect can be exploited by dividing the time domain signals into sections, which are then transformed into the frequency domain. If a section is longer than 5 ms, however, it can lead to 'pre-echo' distortion [4].

4.2.4   Transformation and quantisation

Once the time domain samples have been put into individual sections, each section is transformed into the frequency domain. One method of transformation (used in MPEG-1 Layer II audio [5]) uses polyphase quadrature mirror filters (PQMFs).

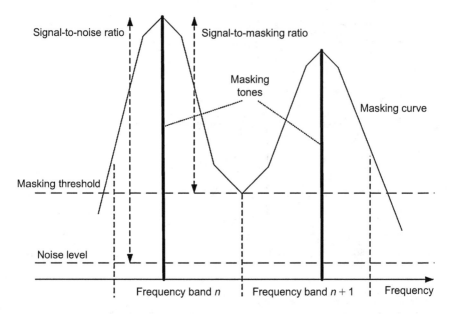

Figure 4.3   Difference between signal-to-noise ratio and signal-to-masking ratio

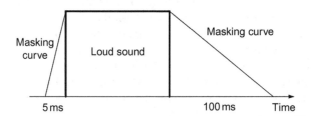

Figure 4.4   Temporal masking effect

These filters transform each section into a number of equally spaced frequency sub-bands. PQMFs have the property that the transformation into the frequency domain can be reversed using equivalent reconstruction filters [6]. The combination of forward and inverse transformations is a loss-less operation, provided that the frequency domain coefficients are not quantised. Bit rate reduction, however, is achieved by quantising the frequency domain coefficients. The combination of transformation and quantisation is one of the main principles of audio as well as video compressions.

The quantisation levels for each frequency band are calculated by a psychoacoustic model, which aims to minimise audible distortion, while achieving the desired bit rate. The model takes into account the frequency response of the ear as well as frequency and temporal masking effects. We will see in the next section that several of the audio compression principles described here are equally applicable to video compression.

4.3   Introduction to video compression

In order to exploit the redundancies in video signals, the characteristics of the human psycho-visual system have to be understood. Many tests have been conducted to measure the response of the visual system in terms of spatial and temporal resolution and sensitivity. For example, spatial resolution was found to be much higher in luma (grey levels) than in chrominance (colour differences). This is because the retina of the human eye contains only about half as many chrominance receptors (cone cells) as luma receptors (rod cells) [7]. Therefore, one first step in compression is to convert the RGB signal into a YCbCr signal and down-sample the number of chrominance samples both horizontally and vertically from a 4:4:4 sampling structure to a 4:2:0 format, as explained in Chapter 2.

In the temporal domain, it is well known that if a series of images with similar content is presented at a rate of 50 images per second or more, it will appear to the human eye as a smooth, continuous motion [8]. Furthermore, the capability of detecting fine detail in moving objects diminishes with the speed of the objects. This trade-off between spatial and temporal resolutions has been exploited in television systems from the very beginning by using interlace techniques, probably the first video 'compression' technique ever used.

In addition to the spatial and temporal frequency responses, there are spatial as well as temporal masking effects, akin to those described in the previous section on audio compression. Whereas in an audio signal a loud tone masks quieter sounds of nearby frequencies, the masking effect in images is due to high-level detail at high spatial frequencies. This is because the eye is less sensitive to detail near objects with high contrast.

There is also an equivalent of the temporal masking effect of audio signals in video sequences before and after scene cuts. Due to the complete change of scenery, it is difficult to perceive details a few frames before and after scene cuts. However, the temporal masking effect is far less pronounced and much shorter in video signals than in audio signals, and it applies only if the cut is instantaneous. Cross-fades between scenes do not have the same effect.

4.4   Redundancy in video signals

Raw (uncompressed) video signals require enormous bit rates or file sizes. An uncompressed digital video signal of standard definition television (SDTV), for example, has a bit rate of 270 Mbit/s. Even after reducing the bit depth from 10 bits to 8 bits and removing all blanking samples, a bit rate of 166 Mbit/s still remains. However, a video signal sampled and digitised on a pixel-by-pixel basis contains considerable amounts of redundant information, i.e. repeated data. These redundancies can be categorised into three groups: spatial, temporal and statistical redundancies.

Spatial redundancy

Spatial redundancy occurs because spatially adjacent pixels within an object are likely to be similar. Although there are, of course, step changes from one object to another, generally there are often large areas in a picture with similar pixels values.

Temporal redundancy

Video signals are captured on a picture-by-picture basis. To give the impression of a smooth motion, 50 pictures (i.e. fields in 625-line TV signals) are captured per second. Therefore, the differences between one picture and the next are often very small.

Statistical redundancy

Once spatial and temporal redundancies have been reduced, the pictures are encoded using a set of symbols for coding entities such as transform coefficients or motion vectors. Statistical redundancy can be exploited by using shorter codewords for more frequent symbols and longer codewords for less frequent ones.

4.5   Exploiting spatial redundancy

There are many different techniques for exploiting spatial redundancy, but most of them use some form of two-dimensional mathematical transformation. Although different algorithms use different transformations, they all have the property of accurately reproducing the original samples once the inverse transform is applied to the transform coefficients. In other words, the combination of transformation and inverse transformation is a loss-less operation. The purpose of the transformation is to differentiate between lower and higher spatial frequencies. The former are generally visually more important and are therefore coded more accurately than the latter.

Most compression algorithms use a two-dimensional, block-based transformation. However, sub-band coding and wavelet algorithms apply the transformation to the entire image (see Chapter 6). For the transformation, the choice of block size is a trade-off between using a block large enough to compact the energy most efficiently but small enough to allow for spatial variation and to keep complexity low. The most popular transformation size is 8×8 pixels.

Some MPEG and other compression algorithms use an 8×8 discrete cosine transform (DCT), originally developed for JPEG still-image compression [9]. Although cosine is a floating-point operation, integer approximations have been developed in order to reduce complexity and processing requirements [10]. Nevertheless, an 8×8 DCT with an accuracy of 8 bits in the spatial domain needs an accuracy of at least 12 bit in the transform domain, in order to return to the original samples after forward and inverse transformations. In other words, the sixty-four 8 bit samples in the spatial domain are transformed into the sixty-four 12 bit coefficients in the (spatial) frequency domain. This corresponds to a 50 per cent increase in bit rate. The transformation itself evidently does not reduce bit rate, quite the opposite.

Figure 4.5 shows how an 8×8 spatial block is transformed into a block of 8×8 frequency coefficients. The top-left coefficient represents the average value of all 64 pixels. Coefficient columns further to the right represent higher and higher horizontal frequencies, whereas coefficient rows towards the bottom of the block represent increasingly higher vertical frequencies. For a mathematical definition of DCT and inverse DCT, see Appendix D.

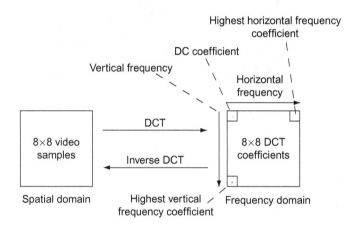

Figure 4.5   Forward and inverse DCT

The spatial redundancy present in the original picture results in a non-uniform distribution of the DCT coefficients. More specifically, in the transform domain, the energy is concentrated towards the lower frequency coefficients. Compression is achieved by quantising the DCT coefficients. In doing that, many coefficients are rounded down to zero after quantisation.

Unlike transformation, quantisation is not a loss-less operation. It introduces a certain amount of quantisation error or quantisation 'noise', and in extreme cases, it can generate blocking artefacts. After DCT, quantisation, inverse quantisation and inverse DCT, we will not get back the exact original pixel values. Instead, we will end up with a block that is slightly different to the original, but hopefully will look visually very similar. Quantisation is performed by dividing each DCT coefficient by a number. Visually important coefficients are divided by a smaller number than less important ones.

4.5.1   Example transformation and quantisation

The processes of transformation and quantisation are best explained by working through an example. Figure 4.6 shows an extract of a standard definition frame taken from a moving sequence. The highlighted block represents a DCT block of 8×8 pixels. Note that the vertical zig-zag contours in the background buildings are not due to some distortion artefact, but are caused by the motion displacement

Figure 4.6   Extract of an SDTV frame with camera pan

between the top and bottom fields. This displacement has a significant effect on the way interlaced blocks are to be coded.

Figure 4.7 shows the highlighted DCT block of Figure 4.6 in numerical form. In this representation, spatial correlation is hard to detect.

Figure 4.8 shows the same block after DCT. Now we can discern a drop in coefficient values towards higher horizontal and vertical frequencies.

Figure 4.9 shows a typical quantisation matrix applied to such a DCT block. Each coefficient in Figure 4.8, with the exception of the top-left DC coefficient,

```
 96  109  125  128  126  124  118  103
111  123  121  113  109  103   92   82
 87   97  119  121  103  103   98   86
 98  114  105   96   96   90   85   83
 93   97  102  106   97   91   91   84
107  115  109  101   93   88   83   82
111  117  119  117  112  105   95   92
120  121  119  116  114  114  105  101
```

Figure 4.7   DCT block in the spatial domain

```
104   393  -318  -83  -112    -4  -14  -16
 39  -110  -149  -34   -56    31   -5   12
476   -61   -63   54    -9     4   12    6
-28   -14     5    3    14    -6  -18    8
 52  -137    22   35     6    -7  -16    9
 51   -69   -42  -23    -9   -15    2    4
 -1   -98   -32  -38   -10    54   19   -7
 50  -176  -139  -35    29   102   21    0
```

Figure 4.8   DCT block in the frequency domain

which is treated separately, is divided by the corresponding number in Figure 4.9 and also by a fixed number (5 in this case). The numerical values of the quantisation matrix correspond to the relative importance of the DCT coefficients in terms of visual picture quality. The higher the number, the less important the corresponding coefficient.

After quantisation, the DCT block consists of just eight coefficients, as shown in Figure 4.10. Note that there is a relatively high content of vertical frequency components as compared to horizontal ones. This is due to the motion between the two fields, as pointed out in Figure 4.6.

```
 8  16  19  22  26  27  29  34
16  16  22  24  27  29  34  37
19  22  26  27  29  34  34  38
22  22  26  27  29  34  37  40
22  26  27  29  32  35  40  48
26  27  29  32  35  40  48  58
26  27  29  34  38  46  56  69
27  29  35  38  46  56  69  83
```

Figure 4.9   Quantisation matrix

```
104   4  -3   0   0   0   0   0
  0  -1  -1   0   0   0   0   0
  5   0   0   0   0   0   0   0
  0   0   0   0   0   0   0   0
  0  -1   0   0   0   0   0   0
  0   0   0   0   0   0   0   0
  0   0   0   0   0   0   0   0
  0  -1   0   0   0   0   0   0
```

Figure 4.10   DCT block after quantisation

4.6   Exploiting statistical redundancy

A quantised block in the transform domain usually contains a large number of zeros, particularly in the higher-frequency regions. By scanning the block from low to high frequencies, we aim to get to the last non-zero coefficient as quickly as possible. Once we have coded the last non-zero coefficient of a block, we can move on to the next block.

Since the quantised matrix contains many zeros, a method of run-level coding has been developed. This combines a number of preceding zero coefficients with the level of the next non-zero coefficient into one codeword. If, for example, two zeros are followed by a $+1$ in scanning order, the three coefficients 0, 0 and $+1$ are combined into one codeword, e.g. 01010 in MPEG-2. Run-level combinations, such as this one, which occur often, are given short codewords, whereas less frequent combinations are given longer variable-length codes (VLC). MPEG-2 defines a total of 112 run-level combinations. The shortest one is the binary number 10, representing no zeros followed by a $+ 1$, and the longest one is the binary number 0000 0000 0001 10111, representing 31 zeros followed by a $- 1$.

However, not all run-level combinations have their own VLC word. In the unusual event that a non-zero coefficient is preceded by more than 31 zeros, for example, the run-level combination has to be coded explicitly. This is done by sending the 'Escape' symbol 0000 01, followed by the number of zeros, followed by the coefficient level. Once the last non-zero coefficient has been run-level VLC coded, an end-of-block (EOB) symbol is sent to indicate that all remaining coefficients of this block are zero.

Following the example discussed in Section 4.5.1, Figure 4.11 shows the order in which the coefficients are scanned with interlaced video in MPEG-2. The grey line indicates the scanning path. This somewhat unusual 'Alternate' scanning order has been developed for interlaced video signals in order to reach the high vertical frequency components as quickly as possible. By comparison, Figure 4.12 shows the more regular zig-zag scan, which is normally used for non-interlaced images.

Using the alternate scanning order, the run-level combinations are coded, as shown in Table 4.1, starting with the first AC coefficient.

In this example, the original block of sixty-four 8 bit pixels (=512 bits) has been compressed to seven run-level codewords, consisting of a total of 50 bits, including the EOB symbol. This is a typical compression ratio for intra-MPEG-2 compression.

```
0   4   6  20  22  36  38  52
1   5   7  21  23  37  39  53
2   8  19  24  34  40  50  54
3   9  18  25  35  41  51  55
10  17  26  30  42  46  56  60
11  16  27  31  43  47  57  61
12  15  28  32  44  48  58  62
13  14  29  33  45  49  59  63
```

Figure 4.11   Alternate scanning order for interlaced blocks

```
0   1   5   6  14  15  27  28
2   4   7  13  16  26  29  42
3   8  12  17  25  30  41  43
9  11  18  24  31  40  44  53
10  19  23  32  39  45  52  54
20  22  33  38  46  51  55  60
21  34  37  47  50  56  59  61
35  36  48  49  57  58  62  63
```

Figure 4.12   Zig-zag scanning order for progressive blocks

Table 4.1   Run-level VLC coding

Run of zeros	Coefficient level	MPEG-2 VLC code
1	5	0000 0001 1011 0
1	4	0000 0011 000
0	−1	111
0	−3	0010 11
0	−1	111
6	−1	0001 011
2	−1	0101 1
EOB		10

4.7   Exploiting temporal redundancy

As mentioned in Section 4.4, in most video clips, there is a high similarity between successive pictures. Typically, most changes are due to objects moving in front of a more or less static background resulting in relatively small areas of the picture being covered and uncovered. Large parts of an image usually stay the same even if they move slightly. This temporal redundancy can be exploited to achieve a significant reduction in bit rate.

The simplest way to exploit this property of video sequences would be to calculate the difference between the picture to be encoded and the previously coded picture. This difference, called the prediction error, uses far fewer coefficients to encode than the picture itself. The method works well for sequences with stationary backgrounds or little motion, but results in large coefficients in sequences containing fast motion – for example a camera pan. In order to reduce the prediction error in sequences with fast motion, a technique called 'motion compensation' is used.

Using motion compensation, the prediction for each block of pixels is not necessarily taken from the same spatial location in the previously coded image. Instead, a search is carried out for each block to find a good match for it somewhere in the previously coded picture. Once a good match has been found, the pixels from that location are copied to the current picture and the error signal is calculated from the motion-compensated block, rather than from the block of the original location. The motion-compensated block is not necessarily the same size as the DCT block. In MPEG-2, for example, motion compensation is carried out on a macroblock-by-macroblock basis, each macroblock consisting of four DCT luma blocks and one DCT block for each colour difference signal. Figure 4.13 illustrates the motion compensation process.

To make sure the decoder can replicate the motion compensation process, the location where the block is taken from has to be transmitted to the decoder in the form of a motion vector. The motion vector describes the position of the prediction block in the previously coded image relative to the position of the block to be coded. Since small displacements are more likely to occur than large ones, small motion vectors are coded with shorter VLC words than large ones.

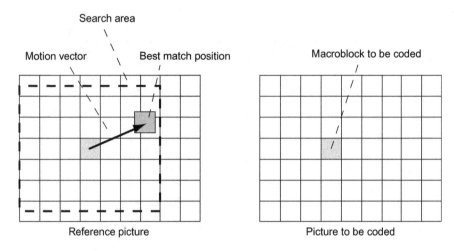

Figure 4.13   Motion compensation of a macroblock

4.8   Motion estimation

A number of different methods have been developed for finding the best prediction block. The most popular techniques are based on block matching. Block-matching techniques use the sum of absolute differences (SAD), calculated over all block pixels, as a prediction criterion. Basic block matching, often referred to as exhaustive motion estimation, is a simple algorithm, but is computationally very expensive [11]. The problem is that a large number of block matches have to be calculated in order to cover a reasonable speed of motion. If the block-matching process is carried out exhaustively, then the SAD has to be calculated for each pixel position over the entire search range. Thus, the computational cost increases with the square of the search range. For this reason, a number of computationally less expensive motion estimation techniques have been developed, as shown in Chapter 7.

4.8.1   Example transformation of a predicted block

Carrying on with the example shown in Section 4.5.1, we now consider a motion-compensated area of the same video sequence. Figure 4.14 shows an extract of the same area but of a motion-compensated predicted frame. The contours shown in the figure are prediction errors. This error signal needs to be (DCT) transformed, quantised, entropy coded and sent to the decoder, together with the motion vectors of the prediction.

Figure 4.15 shows the highlighted DCT block of Figure 4.14 in numerical form. Figure 4.16 shows the same block after DCT.

Figure 4.14   Extract of a predicted frame

```
134  135  130  128  129  134  129  129
124  123  121  123  123  122  126  121
130  125  126  126  127  126  125  126
127  131  134  132  132  132  134  131
127  124  124  127  128  128  129  131
121  121  122  125  127  127  131  133
123  124  123  127  129  131  128  130
127  125  127  126  127  128  126  126
```

Figure 4.15   Predicted DCT block in the spatial domain

```
127  -61    5   24    5    1    0   -1
 25   72   17   13  -18    3   -8    0
-25   45    7   27  -14  -12    8  -15
 18  -51   21    4    2  -13   -7  -16
114   36   -2   -3  -23   -7   12   -2
114   -1  -17   -3  -20  -20   10  -11
 20    9   18   -1    9   -3   16    4
 -8   21   13   24   26   -9   18  -12
```

Figure 4.16   Predicted DCT block in the frequency domain

Although MPEG-2 would allow the use of a quantisation matrix in a predicted frame, a flat quantisation is often used. After quantisation we obtain the quantised DCT coefficients, shown in Figure 4.17.

The run-level combinations shown in Table 4.2 are coded using the alternate scanning order. In this case, the original block of sixty-four 8 bit pixels has been compressed to two VLC words of 11 bits, not including the EOB symbol and motion vector bits.

```
127   0   0   0   0   0   0   0
  0   0   0   0   0   0   0   0
  0   0   0   0   0   0   0   0
  0   0   0   0   0   0   0   0
  1   0   0   0   0   0   0   0
  1   0   0   0   0   0   0   0
  0   0   0   0   0   0   0   0
  0   0   0   0   0   0   0   0
```

Figure 4.17   Predicted DCT block after quantisation

Table 4.2   Run-level VLC coding of predicted block

Run of zeros	Coefficient level	MPEG-2 VLC code
9	1	0000 1011
0	1	110
EOB		10

4.9   Block diagrams

Figure 4.18 shows a block diagram of a generic video encoder. The functionality of most processing blocks has already been explained. As can be seen, the coding loop is preceded by a pre-processing function, which includes functions such as 4:2:2 to 4:2:0 chroma conversion, picture re-sizing and noise reduction, among other functions. More details about the pre-processing functions will be provided in Chapter 8. After entropy coding, the compressed bit stream is stored in an encoder rate buffer before it is transmitted to the decoder. The purpose of this buffer is explained in the following paragraph.

Figure 4.19 shows a block diagram of a generic video decoder for the completion of the video compression codec. The post-processing block contains functions such as 4:2:0 to 4:2:2 chroma conversion, as explained in Chapter 11. Comparing Figure 4.18 with Figure 4.19, it can be seen that the lower part of the encoder loop is essentially the same as the decoder loop, except that the decoder does not need a motion estimator. Having a 'decoder' in the encoder ensures that the predictions in the encoder and decoder are synchronised.

4.10   Quantisation control

Quantisation is the coding tool that allows us to control not only the overall bit rate but also the number of bits allocated to each part. At the lowest level, we can employ quantisation matrices to use different quantisation levels for different DCT coefficients. Lower spatial frequency coefficients are usually less quantised than higher ones to achieve best picture quality. Furthermore, we can use different

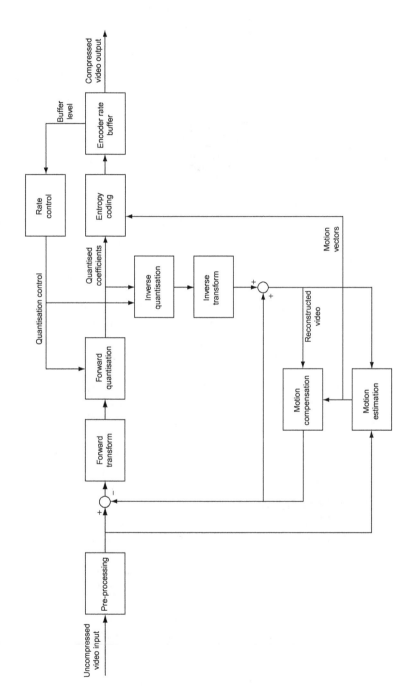

Figure 4.18   Block diagram of a generic video encoder

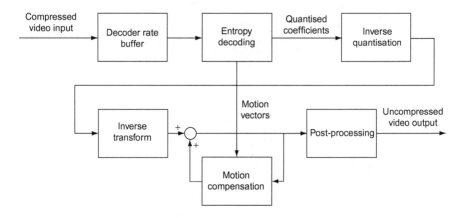

Figure 4.19   Block diagram of a generic video decoder

quantisation matrices for intra- and inter-coded blocks as well as for luma and chroma blocks.

Within each picture we can change quantisation from one macroblock to the next with the aim of allocating more bits to the visually most important picture areas. Picture areas with lower spatial activity are usually more demanding than those with high contrast. This is due to the masking effect mentioned in Section 4.2.

At the highest level we can use different quantisation levels on different picture types. So far we have come across only two picture types: intra-coded pictures (I pictures) and predicted pictures (P pictures). I pictures take no predictions from other pictures and can be decoded without reference to other pictures, whereas P pictures take predictions from previous I or P pictures. A third picture type is introduced in the next section.

4.11   Bi-directional prediction

When objects move in front of a background, there are uncovered areas that cannot be predicted from past pictures because they are hidden by the object. If, however, we had access to a future picture, we could predict the uncovered areas. Of course, we would need to code (and transmit) this 'future' picture before we could make predictions from it. This means we need to make predictions across several pictures in order to obtain a future reference picture from which we can make the predictions. As a result, we need a greater search range because objects will have moved further across several pictures than from one picture to the next. However, it has been found that bi-directional prediction often provides a net coding benefit. Figure 4.20 illustrates the prediction modes for bi-directional (B) pictures.

One of the reasons why B pictures help to save bit rate is that since – in MPEG-2 – no predictions are made from B pictures, they can be more heavily

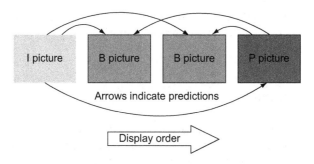

Figure 4.20   B picture predictions

quantised without causing a noticeable picture degradation. Furthermore, B pictures tend to provide coding advantages in scenes with complex motion, for example zooms, slow-moving detailed images and images with noise or film grain.

If B pictures are used, the coding and transmission order is different from the capture and display order, since predictions in the decoder can be made only from pictures that have already been coded and transmitted. Prior to the encoding process, the pictures are reordered in the encoder into transmission order. After decoding the pictures in transmission order, the decoder reorders the pictures back into display order. The difference between display order and transmission order is shown in Figure 4.21. The indices refer to the picture types in capture/display order.

Capture order:
I2 B3 B4 P5 B6 B7 P8 B9 B10 P11 B12 B13 P14 B15 B16

Transmission order:
I2 B0 B1 P5 B3 B4 P8 B6 B7   P11 B9   B10 P14 B12 B13

Figure 4.21   Reordering of B pictures

4.12   Mode decision

Even with the best motion estimation technique, there will be cases in which it is better to intra-code a macroblock rather than use the prediction. In B pictures, there are even more choices available. Apart from intra-coding, predictions can be taken from a past or a future picture, or a combined prediction can be made from a past and a future picture.

The decision as to which macroblock mode should be chosen is not a trivial one. Simple mode decisions compare the level of distortion of each mode and choose the one with the lowest distortion. However, this could require a disproportional number of bits. Ideally, one should consider the number of bits generated by a particular macroblock mode as well as the level of distortion and choose a suitable compromise. This method is called rate-distortion optimisation (RDO) and is an area of active research [12,13].

4.13   Rate control

The DCT coefficients, motion vectors and any additional information required to decode the compressed video signal form a bit stream, which is stored in a buffer before it is sent to the decoder. The buffer, often referred to as 'rate buffer', is necessary to convert the highly variable bit rate generated by the coding loop into a constant bit rate. Figure 4.19 shows the feedback loop from the encoder rate buffer to the forward and inverse quantisation blocks. As we have seen in the examples shown in Section 4.6 and 4.10, intra-coded pictures typically generate many more bits than predicted pictures. However, the output bit rate of an encoder is usually required to provide a constant bit rate. Even if the channel capacity has some headroom, it might not be able to absorb the high bit rates generated during intra-coded pictures. By controlling the quantisation, the rate control algorithm ensures that the encoder rate buffer neither overflows nor underflows.

Figure 4.22 shows how the buffer levels of the encoder and decoder are anti-symmetric. A high buffer level in the encoder rate buffer corresponds to a low buffer level in the decoder rate buffer, and vice versa. Therefore, the total data held in the encoder and decoder rate buffers (in constant bit rate systems) is always constant.

The occupancy of the encoder rate buffer can be used to control the quantisation and inverse quantisation of the encoding loop in order to produce the desired output bit rate. If the encoder rate buffer fills up too much because the video signal is highly complex, the rate control increases quantisation (and inverse quantisation) in order to avoid buffer overflow. If, on the other hand, the video signal is less critical, quantisation can be reduced. In practice, algorithms for rate control are somewhat more sophisticated in order to achieve the best bit allocation within the buffer constraints [14,15].

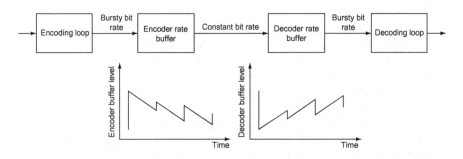

Figure 4.22   Buffer levels of encoder and decoder rate buffers

4.14   Concluding remarks

The explanations and examples discussed in this chapter are largely based on the MPEG-2 standard. Although MPEG-4 (AVC) is rapidly becoming the predominant coding standard, compression principles are somewhat easier to explain in terms of MPEG-2 coding tools and terminology. Nevertheless, the basic ideas introduced here are readily translated and expanded into more advanced coding standards, as outlined in the next two chapters.

4.15   Summary

- Transformation is a loss-less operation. It does not reduce bit rate, but helps to remove spatial redundancy.
- Bit rate reduction is achieved with quantisation and entropy coding, once spatial and temporal redundancies have been removed.
- Quantisation can be varied to achieve the desired bit rate and, depending on masking effects, to achieve optimum picture quality at a given bit rate.
- Temporal redundancy is removed by finding good predictions from previously coded images.

References

1. T. Moscal, *Sound Check: Basics of Sound and Sound Systems*, Milwaukee, WI: Hal Leonard Corporation, ISBN 079353559X, 1995.
2. J. Skoglund, W.B. Kleijn, 'On Time–Frequency Masking in Voiced Speech', *IEEE Transactions on Speech and Audio Processing*, Vol. 8, Issue 4, July 2000. pp. 361–9.
3. F. Sinaga, T.S. Gunawan, E. Ambikairajah, 'Wavelet Packet Based Audio Coding Using Temporal Masking', International Conference on Information, Communications and Signal Processing, Vol. 3, December 2003. pp. 1380–3.
4. M. Bosi, R.E. Goldberg, *Introduction to Digital Audio Coding and Standards*, New York: Springer, ISBN 1402073577, 2002.
5. International Standard ISO/IEC 11172-3 MPEG-1 Audio, 1993.
6. G. Zelniker, F.J. Taylor, *Advanced Digital Signal Processing: Theory and Applications*, New York: Marcel Dekker, ISBN 0824791452, 1993.
7. R.V. Krstic, *Human Microscopic Anatomy: An Atlas for Students of Medicine and Biology*, Springer-Verlag New York Inc., ISBN 0387536663, 1991.
8. M. Sonka, V. Hlavac, R. Boyle, *Image Processing, Analysis, and Machine Vision*, London: Thomson Learning, ISBN 049508252X, 2007.
9. W. Effelsberg, R. Steinmetz, *Video Compression Techniques: From JPEG to Wavelets*, San Francisco: Morgan Kaufmann, ISBN 3920993136, 1999.
10. L.V. Kasperovich, V.F. Babkin, 'Fast Discrete Cosine Transform Approximation for JPEG Image Compression', International Conference on Computer Analysis of Images and Patterns, Budapest, September 1993. pp. 98–104.

11. M. Gharavi-Alkhansari, 'A Fast Motion Estimation Algorithm Equivalent to Exhaustive Search', IEEE International Conference on Acoustics, Speech and Signal Processing, Salt Lake City, May 2001. pp. 1201–4.
12. Z. He, S.K. Mitra, 'A Unified Rate-Distortion Analysis Framework for Transform Coding', *IEEE Transactions on Circuits and Systems for Video Technology*, Vol. 11, Issue 12, December 2001. pp. 1221–36.
13. S. Wan, F. Yang, E. Izquierdo, 'Selection of the Lagrange Multiplier for 3-D Wavelet-Based Scalable Video Coding', International Conference on Visual Information Engineering, London, July 2007. pp. 309–12.
14. Z. Chen, K.N. Hgan, 'Recent Advances in Rate Control for Video Coding', *Signal Processing: Image Communication*, Vol. 22, Issue 1, January 2007. pp. 19–38.
15. S.C. Lim, H.R. Na, Y.L. Lee, 'Rate Control Based on Linear Regression for H.264/MPEG-4 AVC', *Signal Processing: Image Communication*, Vol. 22, Issue 1, January 2007. pp. 39–58.

Chapter 5
MPEG video compression standards

5.1 Introduction

Having investigated the basic principles of video compression, it is time to have a look at some real compression algorithms. By far the most widely used family of video compression algorithms used in broadcast and telecommunication applications are the MPEG algorithms. The Moving Picture Experts Group (MPEG) is a working group of ISO/IEC (the International Organization for Standardisation/International Electrotechnical Commission), i.e. a non-governmental international standards organisation. The first MPEG meeting was held in May 1988 in Ottawa, Canada. To date, MPEG has produced a number of highly successful international standards for video compression, as well as for multimedia content management.

One of the principles of all MPEG standards is that they define coding tools as well as bit-stream syntax and semantics, but they do not specify encoding algorithms. Although MPEG publishes reference software for encoding and decoding, these are merely example programs and are not necessarily achieving the highest picture quality. It is therefore not possible to define an 'MPEG quality' and compare it with a non-MPEG standard.

5.2 MPEG-1

The goal of MPEG-1 [1] was to standardise a video encoding algorithm cap-able of compressing a feature-length movie onto a video CD at SIF resolution (352 pixels by 288 lines, 25 frames/s, or 352 pixels by 240 lines, 29.97 frames/s) [2]. This corresponds to an average bit rate of about 1 Mbit/s, although the maximum bit rate was set to 1.856 Mbit/s (with the *constrained_parameters_ flag* set).

Intra-coding of MPEG-1 is derived from the JPEG (Joint Photographic Experts Group) standard for still image compression, which uses an 8×8 DCT on luma as well as chroma samples. Since the chrominance signals are down-sampled horizontally as well as vertically by a factor of 2 to the 4:2:0 format, there are four luma DCT blocks, corresponding to each one of the two chrominance DCT blocks, as shown in Figure 5.1. Together these six DCT blocks form a macroblock. Therefore, intra-MPEG-1 compression is very similar to JPEG still image compression.

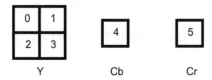

Figure 5.1   Six 8 × 8 DCT blocks form one 16 × 16 macroblock in 4:2:0 format

The 8×8 DCT was chosen for luma as well as chroma transformations because, at the time when the standard was finalised, 8×8 DCT (accelerator) chips were readily available but real-time transformation and encoding in software was not feasible. For a mathematical definition of the two-dimensional 8×8 DCT and inverse DCT, see Appendix D.

In MPEG-1, motion compensation is carried out on a macroblock-by-macroblock basis. This means that there is one motion vector per macroblock describing the motion of all six DCT blocks. The motion vector accuracy in MPEG-1 is 0.5 pixels. This accuracy is needed to provide smooth motion rendition on natural movement that is not pixel aligned.

Furthermore, MPEG-1 includes B frame coding and adaptive quantisation and allows downloading of individual intra- and non-intra-quantisation matrices for each sequence. Bearing in mind that it was the first of four video compression standards known so far, MPEG-1 is quite an efficient algorithm for the compression of non-interlaced (e.g. film-originated) video material.

In addition to I, P and B pictures, MPEG-1 also introduced the concept of D pictures. D pictures contain only low-frequency information, i.e. DC coefficients [3]. They were intended for fast search modes whereby the low-frequency information is sufficient for the user to find the desired video location. Although MPEG-2 should support D pictures to maintain backward compatibility, the concept of D pictures is no longer used in today's video transmission systems.

The ISO/IEC 11172 (MPEG-1) standard consists of five parts, as depicted in Table 5.1.

Although MPEG-1 was intended for movie compression onto video CD at SIF resolution, the algorithm was powerful enough to be used on interlaced SD video signals. With SDTV bit rates of 8–10 Mbit/s, it was more powerful than most compression algorithms available at the time. Moreover, since the algorithm was suitable for real-time hardware implementation, it was immediately used for satellite point-to-point links. This extended use of MPEG-1 was sometimes referred to as MPEG-1.5 – a marketing term that demonstrated that an improved standard for interlaced SDTV signals was urgently needed.

Table 5.1   MPEG-1 structure

Part	Content
1	Systems
2	Video
3	Audio
4	Conformance
5	Reference software

5.3   MPEG-2

Once the compression efficiency of MPEG-1 had been established, it was clear that a similar standard, but one geared towards SD interlaced video signals, would have a wide range of applications, from satellite point-to-point links to direct-to-home (DTH) applications. Today, MPEG-2 [4] is used not only in satellite, cable and digital terrestrial DTH systems but also in many professional applications, such as in contribution links between network studios, in distribution links to transmitters, on file servers and tapes and, of course, on DVDs. Although it is gradually being replaced by MPEG-4 advanced video coding (AVC) (see Section 5.4), it is still used in many applications.

The standardisation process of MPEG-2 was initially focussed on what we now call Main Profile tools, even though profiles were defined much later in the standardisation process. Above all, it was important that real-time encoders and decoders could be implemented cost-effectively while making sure that the highest picture quality could be achieved with a suitable set of coding tools and hardware components.

For example, it was important that an MP@ML decoder could be implemented using a single 16 Mbit synchronous dynamic random access memory (SDRAM) chip. Therefore, the maximum video buffering verifier (VBV) buffer at MP@ML was defined as $1.75 \times 1024 \times 1024 = 1\ 835\ 008$ bits because most of the 16 Mbit capacity was needed for three frame stores to decode bit streams with B frames (two reference frames plus the decoded B frame). It works out that three B frames in 4:2:0 format require $720 \times 576 \times 1.5 \times 8 \times 3 = 14\ 929\ 920$ bits. This leaves $16 \times 1024 \times 1024 - 14\ 929\ 920 = 1\ 847\ 296$ bits for the decoder rate buffer.

This highly focussed approach was one of the main reasons for the enormous success of MPEG-2 because it meant that encoder and decoder chips could be developed in parallel with the standardisation process. The disadvantage of this fast process (MPEG-2 Main Profile was finalised less than 2 years after MPEG-1) was that new coding tools that were still being researched and developed, for example different transformation algorithms, could not be considered because this would have delayed the standardisation process.

Initially, the most important improvement of MPEG-2 over MPEG-1 was that it addressed the compression of interlaced video. Ignoring the spatial offset between the top and the bottom field, interlaced video can be considered as a series of pictures of half the vertical size but with twice the picture rate, and it could easily be coded as such. However, MPEG-1 has shown that interlaced video with relatively slow-moving content (i.e. the majority of video content) is better coded on a frame-by-frame basis. This is because the correlation between frame lines is usually higher than the correlation between field lines, which are spatially further apart. On the other hand, fast-moving material is better treated on a field-by-field basis.

5.3.1 Interlace coding tools

To address this issue, a number of new coding tools were introduced, arguably the most important of which is field–frame DCT coding. Field–frame DCT coding works by treating each macroblock in one of two ways: if the correlation between frame lines is higher than the correlation between field lines, then the vertical transformation is carried out across lines of both fields (frame DCT); otherwise field-based DCT is used. Figure 5.2 gives an illustration of the two modes. The decision is made on a macroblock-by-macroblock basis, so that slow-moving picture areas benefit from frame coding, while fast-moving areas can be field coded. Note that the field–frame decision applies only to luma samples. Chrominance DCT is always frame based.

Similar to the field–frame DCT, macroblocks can be field or frame predicted. This is done by splitting the macroblock into two 16 × 8 blocks, one

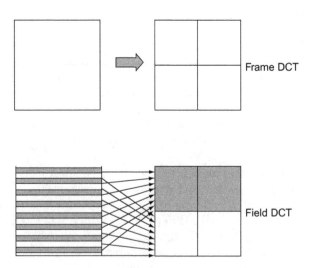

Figure 5.2 Field–frame DCT coding

for each field. If field prediction is chosen, each 16×8 block can take predictions from previous top or bottom fields. Once the predictions are made in field or frame mode, the DCT can be carried out independently in field or frame mode; i.e. it is not tied to the field–frame prediction decision.

If there is fast motion across the entire image, then the coder can change to field-picture coding mode, whereby each field is coded as a separate picture. Although this mode was written into the standard from day one, early consumer decoder chips did not support it.

In addition to the field–frame coding modes, there is a flag in the picture coding extension header of MPEG-2 bit streams that changes the scanning order of DCT coefficients. Despite all field–frame coding modes, interlaced video signals tend to have higher vertical frequency content than progressive signals. Therefore, it would be desirable if high vertical frequency coefficients could be scanned before high horizontal ones. With normal zig-zag scanning, horizontal and vertical frequencies are treated equally. However, if the *alternate_scan* flag is set in the picture coding extension header, then the scanning order is changed, such that high vertical frequency coefficients are reached earlier in scanning order than high horizontal ones. The two scanning orders are shown in Figures 4.11 and 4.12 in the previous chapter.

5.3.2   Film coding tools

Given that MPEG-1 was first and foremost developed for compression of movie material, it perhaps seems surprising that new compression tools for film coding were introduced in MPEG-2 as well. In fact, rather than calling them 'tools', film coding 'flags' would perhaps be a more appropriate description, since most of these flags do not change the compression algorithm itself but supply important information about the video source to the decoder.

There are two flags indicating progressive modes: one in the sequence extension header and one in the picture coding extension header. The former (*progressive_sequence*) signals that the entire sequence is progressively scanned, while the latter (*progressive_frame*) refers only to a single picture. Furthermore, these flags also change the meaning of two further flags: *top_field_first* and *repeat_first_field*. These flags are used to convey the '3:2 pull-down sequence' to the decoder, as shown in Figure 5.3.

Film is usually captured at 24 frames/s. When film is converted to 525-line video at 29.97 frames/s, a method called '3:2 pull-down' is usually applied. To make 30 frames out of 24, every fifth field is repeated. In other words, the telecine converter 'pulls down' three fields from one film frame and two from the next and so on. Since it would be wasteful to re-encode fields that have already been coded and transmitted to the decoder, it is more efficient to tell the decoder to repeat the field after decoding, rather then to re-send the same information. However, if a (repeated) field is dropped in the coding process, the next video frame starts with the bottom field. This has to be indicated to the decoder too, hence the two appropriately named flags. Figure 5.3 illustrates this process.

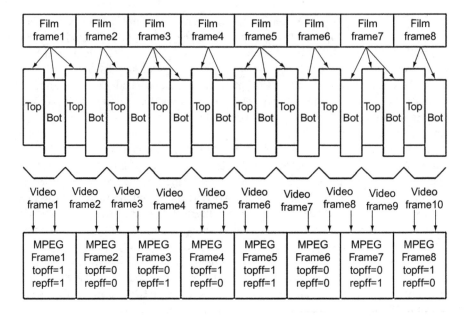

Figure 5.3   3:2 pull-down process and its reversal in MPEG-2 (topff =
top_field_first; repff = repeat_first_field)

This process applies to 525-line SDTV, where film frames with repeated fields have to have the *progressive_frame* flag set. If, however, *progressive_sequence* is set, then the meaning of the *top_field_first* and *repeat_first_field flags* changes completely, as shown in Table 5.2. This second version of 3:2 pull-down is used for converting film sources to 720-line progressive HDTV at 60 or 59.94 frames/s.

Table 5.2   3:2 pull-down flags when progressive_sequence
is set

top_field_first	*repeat_first_field*	**Repeated pictures**
0	0	0
0	1	1
1	1	2
1	0	Forbidden

5.3.3 Low-delay mode

Since MPEG-2 is intended not just for DTH systems but also for live point-to-point links, it introduced the concept of a low-delay mode. There is a low-delay flag in the picture coding extension, which indicates that there are no B frames in the bit stream and B frame reordering is not required. The total end-to-end delay of MPEG systems depends on three main factors: bit rate, rate buffer size and the number of B frames. To minimise the end-to-end delay, the ratio of buffer size/bit rate should be kept as small as possible and B frames should be avoided to eliminate the reordering delay.

There is, of course, a trade-off between end-to-end delay and picture quality. However, in order to improve the performance of P picture coding, MPEG-2 introduced the concept of dual-prime prediction. Dual-prime prediction allows field-based predictions on P frames (from top and bottom fields) using a single motion vector and small differential motion vectors [5]. Dual-prime prediction can be used only if no B frames are present in the bit stream. Although dual-prime predictions provide a modest improvement of coding performance in P pictures, the concept was not carried over to MPEG-4 standards.

5.3.4   Profiles and levels

One of the reasons why MPEG-2 was, and still is, so enormously successful is that its coding tools, bit-rate ranges and video formats for many diverse applications are divided into a number of different compliance points. While MPEG-1 had only one compliance point (if the constrained_parameter_flag is set), MPEG-2 started off with 11. These are divided into a matrix of profiles and levels. Profiles define the decoding capability in terms of coding tools, and levels define constraints on bit rates, frame rates and picture sizes. Table 5.3 gives an overview of the most important MPEG-2 profiles and levels. Note that the SNR, Spatial Scalability and High Profiles have so far not been used in broadcast applications.

The advantage of such a structure is that decoder manufacturers who develop decoder chips for one particular application (e.g. DTH SDTV) can concentrate on one particular compliance point (Main Profile and Main Level in this case) and ignore all other coding tools and picture sizes. Furthermore, the chip can be tested against the limits defined in the level, thus making sure that the chip is interoperable with any encoder working to the same compliance point.

There are four levels in MPEG-2: Low, Main, High-1440 and High. These levels correspond to CIF, SDTV, 4* SDTV and HDTV, respectively. Note that the High-1440 level was originally proposed as an HDTV format and is needed for the Spatial Profile.

The profiles are: Simple, Main, 4:2:2, SNR, Spatial and High. Of these, Main and 4:2:2 Profiles are the most important ones for DTH and professional applications, respectively. Therefore, only four compliance points are relevant to broadcast applications: Main Profile and 4:2:2 Profile, both at Main Level and High Level.

The Simple Profile is the same as the Main Profile, except that it does not support B frames. This was originally proposed for low-delay applications, where B frames are not used. Since Main Profile decoders can decode bit streams with or without B frames, this compliance point is no longer important.

Table 5.3   MPEG-2 profiles and levels

Level		Profile					
		Simple	Main	4:2:2	SNR	Spatial	High
	Picture type	I, P	I, B, P	I, B, P	I, B, P	I, B, P	I, B, P
	Chroma format	4:2:0	4:2:0	4:2:2	4:2:0	4:2:2	4:2:2
High	Samples/line		1920	1920			1920
	Lines/frame		1088	1088			1088
	frames/s		60	60			60
	Bit rate (Mbit/s)		80	300			100
High-1440	Samples/line		1440			1440	1440
	Lines/frame		1088			1088	1088
	frames/s		60			60	60
	Bit rate (Mbit/s)		60			60	60
Main	Samples/line	720	720	720	720		720
	Lines/frame	576	576	608	576		576
	frames/s	30	30	30	30		30
	Bit rate (Mbit/s)	15	15	50	15		20
Low	Samples/line		352		352		
	Lines/frame		288		288		
	frames/s		30		30		
	Bit rate (Mbit/s)		4		4		

SNR and Spatial Profiles are for scalable video coding (SVC). With the SNR method (SNR because of the difference in signal-to-noise ratio between the two layers), a video signal is coded in two bit streams: a low bit-rate stream that is Main Profile compliant and a second one that is used to improve the picture quality and hence the SNR of the same video content. Only SNR-compliant decoders can decode both bit streams and provide the high-quality video signal. A similar approach is used in the Spatial Profile, but this time the two bit streams encode two different resolutions (Main and High-1440 Levels). A Main Profile Main Level decoder can decode the lower resolution bit stream, but a Spatial Profile-compliant decoder is required to combine the two bit streams and display the HDTV signal. We find more on spatial scalability in Chapter 9. The High Profile simply incorporates all MPEG-2 coding tools.

The ISO/IEC 13818 (MPEG-2) standard consists of nine parts, as shown in Table 5.4.

5.3.5   Block diagram

Figure 5.4 shows a basic block diagram of an MPEG-2 encoder. At this level there seem to be no differences between MPEG-1 and MPEG-2 because the

Table 5.4   MPEG-2 structure

Part	Content
1	Systems
2	Video
3	Audio
4	Conformance
5	Reference software
6	Digital storage media, command and control (DSM-CC)
7	Advanced audio coding (AAC)
8	Extension for real-time interfaces
9	Conformance extension for DSM-CC

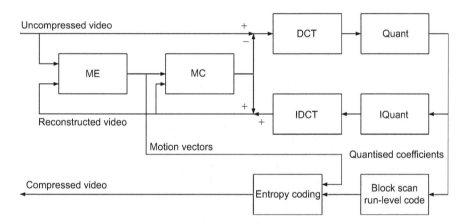

Figure 5.4   Basic block diagram of an MPEG-2 encoder; ME = motion estimation; MC = motion compensation; Quant = quantisation; IQuant = inverse quantisation

field/frame decisions are within the DCT/inverse DCT and motion compensation blocks.

MPEG-3 was originally intended as a standard for HDTV. However, the introduction of profiles and levels in MPEG-2 meant that a separate standard for HDTV was no longer needed. Since the research on video compression continued, a new standardisation process was initiated to overcome some of the limitations of MPEG-2.

5.4   MPEG-4

There are two video compression algorithms standardised in MPEG-4: MPEG-4 Part 2 (Visual) [6] and MPEG-4 Part 10 (AVC) [7]. The two algorithms are completely different. This is not unusual within MPEG standards. For example, there are two algorithms for audio compression in MPEG-2: Part 3 Audio and Part 7 Advanced Audio Coding (AAC). These are also two completely different algorithms.

At the time of writing, MPEG-4 (ISO/IEC 14496) consisted of 23 parts, as shown in Table 5.5.

Table 5.5   MPEG-4 structure

Part	Content
1	Systems
2	Visual
3	Audio
4	Conformance
5	Reference software
6	Delivery multimedia integration framework
7	Optimised reference software
8	Carriage on IP networks
9	Reference hardware
10	Advanced video coding (AVC)
11	Scene description and application engine
12	ISO base media file format
13	Intellectual property management and protection
14	MPEG-4 file format
15	AVC file format
16	Animation framework eXtension
17	Timed text subtitle format
18	Font compression and streaming
19	Synthesised texture stream
20	Lightweight scene representation
21	MPEG-J graphical framework eXtension
22	Open font format specification
23	Symbolic music representation

5.4.1   MPEG-4 Part 2 (Visual)

Since MPEG-1 was designed for SIF resolution, MPEG-2 concentrated on picture sizes of SD and HD. On small, non-interlaced images, MPEG-2 is actually not quite as efficient as MPEG-1. This is because in MPEG-2 a new slice header is mandatory for each row of macroblocks and MPEG-2 interlace coding tools are of no benefit on non-interlaced sequences. Multiple slice headers have the advantage that the decoder can quickly recover from bit-stream errors, but they are not very efficient for images smaller than SD. On the whole, MPEG-2 is quite verbose with its headers. The simplest way of increasing the coding efficiency, particularly for smaller images, is to reduce the number and size of headers. Therefore, the rule that each row of macroblocks has to have its own slice header has been dropped again in MPEG-4 (Visual and AVC).

5.4.1.1   Motion prediction

Initially, MPEG-4 (Visual) focussed on smaller (sub-SDTV) non-interlaced images. With smaller images, two factors needed to be considered first and foremost: the motion compensation block size of MPEG-2 (16 × 16 for non-interlaced images) is too big, and the motion vector precision of MPEG-2 (½ pixel) is not accurate enough.

With smaller block sizes, prediction errors at object boundaries can be reduced. This is illustrated in Figure 5.5. With 16 × 16 motion compensation, Object A would be motion compensated together with the background motion. However, using 8 × 8 blocks, the bottom-left partition can be motion compensated in accordance with the object motion, whereas the other three partitions can be compensated with the background motion, thus reducing the prediction error considerably.

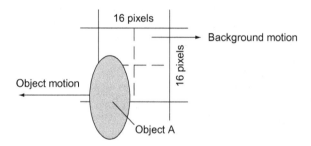

Figure 5.5   Reducing the prediction error at object boundaries

In the same way that natural objects are not macroblock aligned, the movement of natural objects is not pixel aligned. Assuming that a ½ pixel motion vector is accurate enough for SDTV images, CIF images would need a ¼ pixel motion vector precision in order to achieve an equivalent prediction accuracy. By reducing the minimum motion compensation block size to 8 × 8 (16 × 16 motion compensation is also still allowed as an option) and increasing the motion vector accuracy to ¼ pixel, the coding efficiency of MPEG-4 (Visual) is noticeably improved.

Furthermore, MPEG-4 (Visual) allows motion vectors beyond picture boundaries. It seems strange, at first, that predictions from outside the image would improve coding quality. After all, there is no new information available for better predictions. However, the point is that if (on slow motion) only a couple of pixels are outside the image, it makes sense to allow predictions of the rest of the macroblock. This is particularly important for small images, where the percentage of border macroblocks is higher than on larger images. For example, the percentage of border macroblocks increases from 10 per cent for SDTV to 19 per cent for CIF images.

There are two more coding tools relating to predicted frames that were introduced in MPEG-4 (Visual): direct mode prediction in B frames [8] and global motion vectors [9,10]. While the benefit of the latter can be debated (global motion vectors were dropped again in Part 10), direct mode is a highly efficient prediction mode in B frames. In direct mode, the motion vectors for B frame macroblocks are calculated from the motion vectors of the co-located macroblock in surrounding P frames, permitting motion-compensated prediction without having to send motion vectors [11].

5.4.1.2   Intra-prediction

Further to the improvements on predicted frames, MPEG-4 (Visual) also introduced intra-AC predictions in the frequency domain [12]. The idea is that neighbouring macroblocks are often quite similar in appearance. Therefore, once a DCT to the left or above has been coded, intra-predictions can be taken from these blocks. For example, if a vertical edge runs through two vertically adjacent DCT blocks, then the AC coefficients on the top row of the two DCT blocks will be very similar, if not identical. Predicting these coefficients from the DCT block above can reduce the number of residual coefficients considerably. The same applies to horizontal edges with predictions from the DCT block to the left.

In MPEG-4 (Visual), the following profiles have been defined: Simple, Advanced Simple, Simple Scalable, Core and Main. Of these, the Advanced Simple Profile is one of the most widely used for streaming and mobile applications. Furthermore, MPEG-4 (Visual) defined the following levels: Level 1 QCIF, Level 2 CIF, Level 3 SDTV and Level 4 HDTV.

5.4.1.3   Object coding

Apart from compression improvements, MPEG-4 (Visual) also defines methods for binary shape coding (Core Profile). Instead of coding rectangular video images, binary shape coding makes it possible to code arbitrarily shaped individual video objects [13]. Therefore, instead of referring to sequences and pictures, MPEG-4 (Visual) refers to video object layers (VOL) and video object planes (VOP), respectively. Once decoded, these objects are composited at the decoder, and they can be interactively moved by the user to different positions on the screen. The concept of object coding can also be extended to audio signals, where musical instruments can be individually coded and superimposed in the decoder.

However, the concept of arbitrarily shaped object coding has, so far, not been implemented in commercial broadcast projects. In fact, apart from some applications for video streaming over the Internet or to mobile devices, MPEG-4 (Visual) has not been used for broadcast applications and is now all but superseded by MPEG-4 (AVC). The concept of video object coding has not been adopted in MPEG-4 (AVC).

5.4.2   MPEG-4 Part 10 (AVC)

Even before MPEG-4 (Visual) had been finalised, an even more advanced compression algorithm was being developed. Originally known as H.26L, this new algorithm is, in many respects, quite different to MPEG-2 and MPEG-4 (Visual). The two, perhaps most significant, differences are a new integer 4×4 transform instead of the 8×8 DCT and a new entropy-coding algorithm based on arithmetic coding rather than VLC. In order to broaden the algorithm to make it suitable for interlaced video among other things and to integrate it into the MPEG family, a Joint Video Team (JVT) was formed between the ITU and ISO/IEC.

5.4.2.1 New transforms

One of the disadvantages of the 8×8 DCT is that it is inherently a floating-point operation and can therefore be approximated in different ways in an integer format. Different implementations can give slightly different answers due to small rounding errors. This can lead to discrepancies between the inverse DCT in the encoder and the inverse DCT in the decoder. Neither MPEG-1 nor MPEG-2 nor MPEG-4 (Visual) standardised an exact inverse DCT in integer format. Instead, they refer to an IEEE standard on DCT accuracy [14]. The IEEE standard defines the required accuracy of the inverse DCT, such that this mismatch is not visible after several generations of predictions from predictions. Yet this mismatch gradually increases from one prediction to the next and can lead to a 'decoder drift'. This drift builds up until the drift error is cleared with the next I picture. If the decoder drift builds up too much on long groups of pictures (GOPs), it can result in an annoying noise pattern, which increases periodically towards the end of each GOP. The problem is even worse if no B frames are used because the decoder drift builds up more rapidly with each P frame. Consequently, the GOP length of MPEG-1, MPEG-2 and MPEG-4 (Visual) should be kept relatively short.

This problem has been overcome in MPEG-4 (AVC) by defining an integer transform. In fact, as can be seen in Figure 5.6, the 4×4 transform of MPEG-4 (AVC) Main Profile is so simple that it can be implemented without any multiplications, i.e. using only binary shifts and additions [15]. Since the inverse transform operations in the encoder and the decoder are identical, there is no decoder drift. Theoretically, intra-coded refresh frames are not needed in MPEG-4 (AVC). In practice, intra-pictures are inserted at regular intervals to keep channel change time at an acceptable level.

A 4×4 transform is highly efficient for small images with high detail. It is less efficient on relatively plain areas. For this reason, the MPEG-4 (AVC) High Profile also defines an 8×8 integer transform, as shown in Figure 5.7 [16]. This

$$
\begin{matrix}
1 & 1 & 1 & 1 \\
2 & 1 & -1 & -2 \\
2 & -1 & -1 & 1 \\
1 & -2 & 2 & -1
\end{matrix}
$$

Figure 5.6 4×4 transform kernel for MPEG-4 (AVC) Main Profile

```
 8    8    8    8    8    8    8    8
12   10    6    3   -3   -6  -10  -12
 8    4   -4   -8   -8   -4    4    8
10   -3  -12   -6    6   12    3  -10
 8   -8   -8    8    8   -8   -8    8
 6  -12    3   10  -10   -3   12   -6
 4   -8    8   -4   -4    8   -8    4
 3   -6   10  -12   12  -10    6   -3
```

Figure 5.7   8 × 8 transform kernel for MPEG-4 (AVC) High Profile

transform requires multiplications, but it also does not give rise to decoder drift because it involves only integer operations.

Interestingly, there is now a proposal to define an 8 × 8 integer transform for MPEG-2 too. This would eliminate decoder drift in MPEG-2 systems, provided that the new transform is used in both encoder and decoder.

5.4.2.2   Entropy coding

The second major difference between MPEG-4 (AVC) and all previous versions of MPEG is in the way motion vectors and transform coefficients are coded. Whereas MPEG-1, MPEG-2 and MPEG-4 (Visual) use fixed VLC tables, which are 'hard coded' into the standard, MPEG-4 (AVC) uses either context-adaptive VLC (CAVLC) tables or context-adaptive binary arithmetic coding (CABAC) for entropy coding.

The VLC tables for MPEG-2 have been optimised, based on a number of test sequences. Although they work remarkably well for most picture material over a wide range of bit rates, they are sub-optimum for non-typical sequences or extremely low or high bit rates. CAVLC uses different VLC tables, depending on the context. Since neighbouring blocks usually have similar properties, CAVLC selects different VLC tables, depending on the properties of neighbouring blocks. This improves the coding efficiency of the entropy-coding process.

CABAC is also context adaptive, but it uses binary arithmetic coding instead of VLC tables. Arithmetic coding is a process that calculates the coded bits based on a set of coefficients (or motion vectors), rather than each coefficient individually. Therefore, it can cope with highly skewed probability distributions. Whereas VLC tables cannot code probabilities of more than 50 per cent or extremely low probabilities very efficiently, CABAC can allocate the right number of bits to a codeword of almost any probability. For example, CABAC can code a highly likely codeword with less than 1 bit. This cannot be achieved with VLC tables. For a more detailed explanation of arithmetic coding in general or CABAC, see References 17 and 18.

5.4.2.3   Intra-predictions

Intra-predictions in MPEG-4 (Visual) operate in the DCT domain. Since the predictions are limited to $C_{0,x}$ and $C_{x,0}$ coefficients, the predictions are purely horizontal or vertical. Diagonal structures cannot be intra-predicted in MPEG-4 (Visual). To overcome this limitation, intra-predictions in MPEG-4 (AVC) are carried out in the spatial domain. There are a total of nine intra-prediction modes in MPEG-4 (AVC), as shown in Figure 5.8. This makes it possible to make predictions in horizontal and vertical directions, as well as several diagonal directions. These intra-predictions can be applied both to 4×4 blocks and to 8×8 blocks (in High Profile). Note that mode 2 is missing in Figure 5.8. This is because mode 2 is a prediction of the average block value.

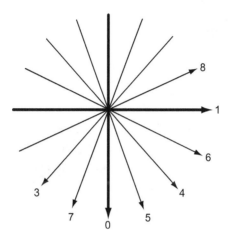

Figure 5.8   Intra-prediction modes in MPEG-4 (AVC)

Further to the intra-prediction modes on 4×4 and 8×8 blocks, it is possible to make the prediction on the entire 16×16 macroblock as a whole. However, only modes 0, 1, 2 and 4 are allowed in 16×16 intra-predictions.

In addition to the 8×8 transforms and intra-prediction modes, MPEG-4 (AVC) High Profile also supports user-defined scaling lists. These are equivalent to the quantisation matrices in MPEG-2.

5.4.2.4   Motion prediction

In addition to the motion-prediction methods introduced in MPEG-4 (Visual), MPEG-4 (AVC) supports a number of new motion compensation modes. In particular, the motion compensation block size can be reduced down to 8×4, to 4×8 and even to 4×4 blocks. All in all, MPEG-4 (AVC) allows a total of seven different prediction block sizes, including 16×16, 16×8, 8×16 and 8×8. However, one has to be careful with small motion compensation block sizes because, although they can provide better predictions in detailed areas with complex motion, they also need more bits to code the additional motion vectors. This could cost more bits than a slightly worse prediction with fewer vectors.

In MPEG-2 and MPEG-4 (Visual), predictions were only possible from I or P pictures. Therefore, the optimum distance between reference (I or P) frames for most sequences was 3, i.e. IBBP coding. However, some sequences were better coded in IBP or IP mode, which is why some encoders use adaptive GOP structures. Even very easy-to-code sequences (e.g. head and shoulder clips) do not benefit much from a higher number of B frames in MPEG-2. MPEG-4 (AVC), on the other hand, allows predictions from B frames. This makes it possible to configure a hierarchical GOP structure, such as shown in Figure 5.9. As can be seen in the figure, once the reference B frame has been coded using predictions from I and P pictures, non-reference B frames can take predictions from neighbouring I, P and/or reference B frames. Higher hierarchical structures, such as those shown in Figure 5.10, are also possible [19].

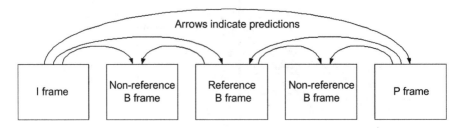

Figure 5.9   Hierarchical GOP structure

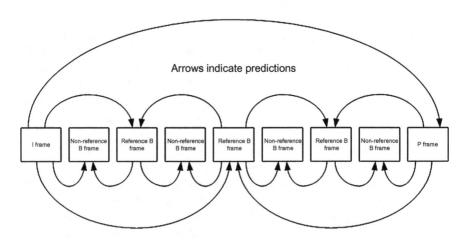

Figure 5.10   GOP structure with seven hierarchical B frames

One of the reasons why a hierarchical GOP structure provides high coding advantages is the use of additional prediction modes in MPEG-4 (AVC)

B frames. Not only is it possible to use forward, backward and combined predictions from past and future reference frames, but MPEG-4 (AVC) also provides direct mode predictions (as did MPEG-4 (Visual)), which make it possible to code non-referenced B frames very efficiently.

In addition to the direct mode introduced in MPEG-4 (Visual), MPEG-4 (AVC) extended the functionality of the direct mode, which is now referred to as 'temporal direct mode', and added a second 'spatial direct mode'. Instead of predicting the motion from co-located temporally neighbouring macroblocks, the spatial direct mode uses the motion vectors from spatially neighbouring macroblocks (above and to the left) to predict the motion of the current macroblock. The high efficiency of direct mode predictions is because in direct mode, motion vectors are predicted rather than coded; i.e. no motion vector bits are transmitted for macroblocks coded in direct mode. This works particularly well in hierarchical GOP structures where predictions can be taken from adjacent (P or reference B) frames.

Further to the prediction modes described in the previous sections, MPEG-4 (AVC) provides two additional modes for temporal prediction: long-term reference pictures [20] and weighted prediction [21]. Long-term reference pictures are kept in the decoder buffer until an 'erase' signal or an IDR (instantaneous decoding refresh) picture is sent. This means that predictions can be made from pictures transmitted a long time ago. Weighted predictions allow an encoder to specify the use of scaling and offset in motion prediction. Implicit weighted prediction for B frames can provide significant benefits during fades.

5.4.2.5   In-loop filter

Despite the high compression efficiency of MPEG-4 (AVC), it is still possible to reduce the bit rate to a point at which the picture information cannot be fully coded any more. At that point the encoder has to eliminate most AC transform coefficients. This leads to the blocking artefacts similar to those seen in critical MPEG-2 sequences. To mitigate this drawback, MPEG-4 (Visual) recommends a post-processing filter that would clean up the picture after decoding. However, since it is not part of the decoding process itself, it was not included in the standard and is therefore not mandatory.

MPEG-4 (AVC), on the other hand, specifies a de-blocking filter operating on the reconstructed image within the encoder and the decoder. Therefore, it is an integral part of the encoding and decoding process and has to be supported by compliant decoders. Furthermore, it has been shown that in-loop de-blocking filters are more efficient in removing blocking artefacts than post-processing filters [22].

Instead of degrading into blocking artefacts, the in-loop filter adaptively removes the block edges so that the picture appears slightly softer, i.e. with less detail. This is visually much more acceptable than blocking artefacts. Figures 5.11 and 5.12 demonstrate compression artefacts of MPEG-2 and MPEG-4 (AVC),

*Figure 5.11   Extract of an SDTV sequence compressed to 1 Mbit/s
in MPEG-2*

*Figure 5.12   Extract of an SDTV sequence compressed to 250 kbit/s in
MPEG-4 (AVC)*

respectively. It can be seen that MPEG-2 degrades into blocking artefacts
whereas MPEG-4 (AVC) loses some detail but is visually more acceptable. Note
that the bit rate of MPEG-4 (AVC) is considerably lower than that of MPEG-2
in order to demonstrate the coding artefacts.

Whereas all other coding tools try to remove redundant information, the
in-loop filter operates on images where the removal of redundant information
is not enough to achieve the desired bit rate.

5.4.2.6   Rate-distortion optimisation (RDO)

RDO is not a coding tool specific to MPEG-4 (AVC). It is a method for choosing the best coding mode in any compression algorithm. The reason why we are listing it as an MPEG-4 (AVC) coding tool is that it is arguably more important in MPEG-4 (AVC) than in any previous MPEG compression standard [23].

Consider two macroblock modes: one mode, e.g. intra-coding, produces low distortion but needs a relatively high number of bits to code, whereas the other mode, e.g. predictive coding, produces a higher distortion but with fewer bits. Which one should be chosen? RDO takes both distortion and bit cost into account by calculating an overall cost function. This is done by converting the bit cost through a Lagrangian operator into a 'distortion equivalent' cost [24]. By comparing distortions and the distortion-equivalent bit costs of the two macroblock modes, the better mode can be chosen. The higher the number of macroblock modes, the more important the choice of optimum coding mode for each macroblock.

5.4.2.7   Block diagram

Figure 5.13 shows a basic block diagram of an MPEG-4 (AVC) encoder. Despite the fact that MPEG-4 (AVC) is significantly more complex than MPEG-2, the main differences in terms of a basic block diagram are the in-loop filter and the intra-prediction blocks.

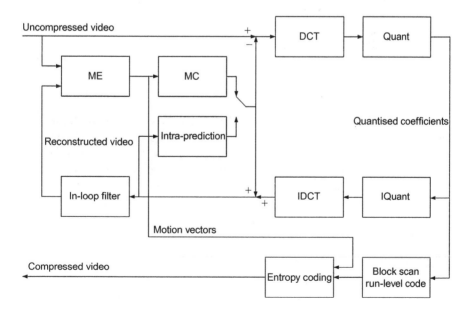

Figure 5.13   Basic block diagram of an MPEG-4 (AVC) encoder

Although it is notoriously difficult to give bit-rate estimations for different applications and coding algorithms, Table 5.6 gives an overview of bit-rate requirements of MPEG-4 (AVC) as compared to MPEG-2. If statistical multiplexing is used, the average bit rate can be reduced further, as shown in Chapter 12.

Table 5.6   Typical bit rates for different applications

Application	MPEG-2	MPEG-4 (AVC)
4:2:2 HDTV contribution and distribution	25..80 Mbit/s	10..50 Mbit/s
4:2:2 SDTV contribution and distribution	15..25 Mbit/s	5..15 Mbit/s
4:2:0 HDTV direct-to-home	10..20 Mbit/s	5..10 Mbit/s
4:2:0 SDTV direct-to-home	3..5 Mbit/s	1.5..2.5 Mbit/s

5.5   Structure of MPEG bit streams

5.5.1   Elementary streams

Having introduced the compression algorithms of MPEG standards, it is worth having a brief look at the way the bit streams are structured. Since the bit-stream structures and timing synchronisation of all MPEG video standards are quite similar, it is best to describe them all in one section. Starting with MPEG-2, Figure 5.14 shows the hierarchy of layers in an MPEG-2 elementary stream from sequence layer down to block layer.

5.5.1.1   Sequence

A sequence is the highest syntactic structure of the coded video elementary stream and consists of several headers and one or more coded pictures. The sequence header contains information about the picture size, aspect ratio, frame rate and bit rate.

The presence of a *sequence_extension* header signifies an MPEG-2 bit stream. The *sequence_extension* header contains the profile and level indicator and the *progressive_sequence* flag. In MPEG-2, sequence header information can only be changed after a *sequence_end_code* has been inserted. Repeat sequence headers must keep the same picture size, aspect ratio, frame rate, profile and level.

Sequence extension headers can be followed by a *sequence_display_extension* containing information about colour primaries and a *sequence_scalable_extension* for scalable bit streams.

5.5.1.2   Group of pictures

A GOP can share a GOP header, which indicates a random access point because the first picture after a GOP header is always intra-coded. The GOP header has information about the time code and signals, whether or not it is a closed GOP. Closed GOPs do not take predictions from previous GOPs.

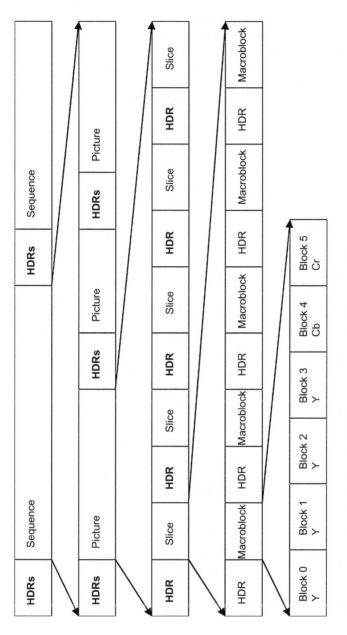

Figure 5.14 Structure of an MPEG-2 elementary bit stream

5.5.1.3   Picture

A picture consists of several headers and a number of slices. In MPEG-2, every row of macroblocks must start with a slice header. Video frames can be coded as one frame picture or as two field pictures. The picture header conveys information about the display order of the pictures, the *picture_coding_type* (I, P or B picture), motion vector sizes and decoding delay. The decoding delay indicates how long a decoder must wait before a picture can be removed from its buffer.

If the sequence header is followed by a *sequence_display_extension*, then the picture header must be followed by a *picture_coding_extension*. This header contains further information about the MPEG-2 bit stream, as shown in Table 5.7.

Table 5.7   Picture_coding_extension flags

Picture_coding_extension flag	Information
F_code	Signifies the maximum motion vector range
Intra_dc_precision	Signifies the number of bits used for DC coefficients, i.e. 8, 9, 10 or 11 bits
Picture_structure	Tells the decoder whether it is a top field, a bottom field or a frame picture
Top_field_first	If set, this flag signifies the first field of a field-coded video frame
Frame_pred_frame_dct	If set, this flag signifies that neither field predictions nor field DCT coding is used in this picture
Concealment_motion_vectors	If set, this flag tells the decoder that intra-coded macroblocks also contain motion vectors
Q_scale_type	If set, the MPEG-2 *quantiser_scale_code* is used; otherwise the MPEG-1 *quantiser_scale_code* is used
Intra_vlc_format	If set, the MPEG-2 intra-VLC table is used; otherwise MPEG-1 intra-VLC table is used
Alternate_scan	If set, alternate block scanning is used; otherwise zig-zag scanning is used
Repeat_first_field	If set, this flag tells the decoder to display the first field of the current frame twice
Chroma_420_type	This flag tells the decoder how to convert from the 4:2:0 format back to 4:2:2
Progressive_frame	This flag tells the decoder that the current frame is not interlaced
Composite_display_flag	If set, this flag indicates that the video source came from a composite video signal, in which case the flag is followed by composite parameters

5.5.1.4   Slice

A picture is coded as a number of slices. A slice consists of one or more macroblocks, all located in the same horizontal row (in MPEG-2). A slice is the smallest MPEG unit with a unique, byte-aligned start code. Therefore, decoding can start (or re-start after an error) at slice level. MPEG video start codes consist of 23 zeros followed by a 1. The slice header also contains the *quantiser_scale_code*.

5.5.1.5   Macroblock

A macroblock describes a picture area of 16×16 luma pixels. Each macro-block consists of a macroblock header, up to four luma blocks and up to two chroma blocks in 4:2:0 (or up to four chrominance blocks in 4:2:2). The macroblock header contains motion vectors, the macroblock coding mode and information as to which, if any, of the blocks are coded. If none of the DCT blocks are coded and there are no motion vectors to be coded, the macroblock is skipped.

5.5.1.6   Block

A block consists of the VLC-coded DCT coefficients, corresponding to an 8×8 array of luma or chroma samples in the spatial domain.

5.5.1.7   Differences in MPEG-4 (AVC)

While the bit-stream structure of MPEG-4 (AVC) is quite similar, there are a number of important differences. Instead of sequence headers and picture headers, MPEG-4 (AVC) uses sequence parameter sets (SPS) and picture parameter sets (PPS). More than one sequence or picture parameter set can be active at any given time, and the slice header indicates which parameter sets are to be used. The picture parameter set contains a section on video usability information (VUI), to convey the aspect ratio and other display information. For example, the vertical position of chroma samples in 4:2:0 mode can be defined in VUI. The decoder can then make sure that the up-sampled chroma samples are at the correct vertical position. This gives more flexibility than in MPEG-2 where the vertical chroma positions were recommended as shown in Figure 2.7 but were not part of the specification.

Whereas in MPEG-2, elementary stream start codes are unique, start code emulation cannot be completely avoided in MPEG-4 (AVC), mainly because of CABAC coding. Therefore, MPEG-4 (AVC) inserts a network adaptation layer (NAL), which removes emulated start codes.

In order to avoid spending too many bits on header information, MPEG-4 (AVC) uses VLC-coded header information and allows slices to cover an entire picture. Furthermore, the core syntax in MPEG-4 (AVC) contains no explicit timing information, such as frame rate, repeated fields, decoder delay, etc. This information is contained in SEI (supplementary enhancement information) messages.

There are two slice types in MPEG-4 (AVC): IDR slices and non-IDR slices. IDR slices define a unique starting point for decoders. No predictions can be made across IDR slices, and all information from previous slices is discarded. While IDR slices have to be intra-coded, non-IDR slices can be intra-coded, forward predicted or bi-directionally predicted.

5.5.2   Transport stream

Part 1 of every MPEG video standard defines the system layer, i.e. the way in which video, audio and other data streams are combined into one bit stream and put on file or transported from one place to another. Part 1 of MPEG-2, and in particular the transport stream (TS) definition, is probably the most widely used MPEG standard created so far. Although created for MPEG-2, it is still in use today (and will be for some time) for MPEG-4 (AVC) transmissions.

MPEG-2 Part 1 describes two ways of multiplexing video, audio and other data streams: using program streams and TSs. Program streams are used for file applications such as on DVDs, whereas TSs are used for transmission applications, including broadcast, point-to-point and even IPTV streaming applications.

A TS consists of packets of 188 bytes, four of which contain header information; the rest is payload data. The header contains information about the payload, as shown in Table 5.8.

Table 5.8   Transport stream header information

Header bits	Information
Sync byte	Each transport stream packet starts with the same sync byte: $71 = 47$ (hex)
Transport_error_indicator	If set, this transport stream packet is likely to contain bit errors
Payload_unit_start_indicator	If set, this transport stream packet contains the start of an access unit, e.g. a video picture or audio frame
Transport_priority	This bit can be used to distinguish between high- and lower priority payloads
PID	This is a 13 bit packet ID number, which identifies each bit stream in the multiplex
Transport_scrambling_control	This indicates whether the payload is scrambled
Adaptation_field_control	This indicates whether an adaptation field is present in the payload
Continuity_counter	A 4 bit counter that increments with every packet of the same PID

Every elementary data stream in the TS multiplex has a unique packet ID number (PID). The standard defines a number of reserved PIDs. PID 0 is reserved for the program association table (PAT) and PID 1 for the conditional access table (CAT). PIDs 2–15 are reserved for future applications, and PID 8191 is reserved for null packets. All other PIDs can be used for video, audio or data streams (as far as MPEG-2 is concerned), although there are other standards, such as ATSC (Advanced Television Systems Committee), which put further constraints on PID numbers.

5.5.2.1 Timing synchronisation

For timing synchronisation in TSs, MPEG-2 Part 1 defines a layer in between the elementary stream and the TS, called the packetised elementary stream (PES). Among other pieces of information, PES packet headers contain the *stream_id*, the *PES_packet_length* and presentation time stamps (PTS). The *stream_id* identifies the content of the PES packet: MPEG-2 video, MPEG-4 (AVC) video, MPEG-1 audio, MPEG-4 (AAC) audio, etc. The PES packet length is used to find the next PES packet header in audio and data streams. In video streams the PES packet length is undefined and the start of the next PES packet is found by examining the *payload_unit_start_indicator* in the transport packet header.

The PTS are used to synchronise video, audio and time-stamped data streams (for example closed captions or subtitles) to the program clock reference (PCR). If B frames are present in the video stream, then two sets of time stamps are provided: decode time stamps (DTS) and PTS. The former defines the time when the access unit, for example the coded picture, is removed from the decoder buffer, whereas the latter defines the time when the decoded picture is presented at the video output. Figure 5.15 shows a simplified timing diagram of a video and audio PES stream with interspersed PCR packets. The PCR packets provide a timing reference in the decoder.

It can be seen in Figure 5.15 that audio access units (audio frames) are presented shortly after the access unit has arrived at the decoder. Some video access units, on the other hand, are not presented until some considerable time after their arrival at the decoder. This is because some video access units, for example I frames, are much bigger than others. Therefore, they take longer to be transmitted using a constant bit rate. However, at the output of the decoder, video frames have to be presented at regular intervals.

5.5.2.2 Adaptation fields

Figure 5.15 shows two PES: video and audio, plus a timing reference in terms of PCR time stamps. However, it does not explain how PCR time stamps are sent to the decoder. The answer is that PCR time stamps are carried in adaptation fields. Adaptation fields can use part of, or the entire payload of, a TS packet. Apart from carrying PCR time stamps, adaptation fields can be used for a number of other purposes, including splice point indicators, random access indicators and TS stuffing.

PCR time stamps are signalled with a higher resolution than PTS and DTS time stamps. Whereas PTS/DTS time stamps are accurate to about 11 µs, PCR time stamps are accurate to within 37 ns. This is necessary because decoders need to reproduce an accurate timing reference that is frequency locked to the encoder timing reference, in order to synchronise the decoding process to the encoder. Otherwise the decoder buffer would eventually underflow or overflow.

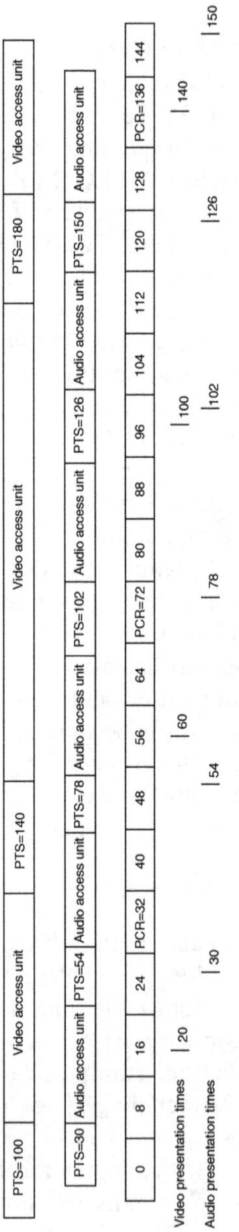

Figure 5.15   Timing diagram of video, audio and PCR bit streams

5.5.2.3 Program-specific information (PSI)

Since video and audio streams can have arbitrary PIDs in a multiplexed TS, there must be some information about the programs the TS contains and, more specifically, about the components the programs consist of and the corresponding PIDs. We have already mentioned in Section 5.7.2 that there are two PIDs that are reserved for specific purposes: PID 0, reserved for the PAT, and PID 1, reserved for the CAT. If a TS contains scrambled bit streams, a CAT must be present in the TS. Furthermore, MPEG-2 defines a network information table (NIT), but it does not specify the content of the CAT and NIT.

The PAT is a table that lists all programs contained in the TS and identifies the PIDs of the program map tables (PMT) of each program. The PMT, in turn, list all elementary streams associated with the program and their PIDs. Furthermore, the PMT provide metadata, such as the video and audio compression formats and descriptors about the bit streams.

MPEG-2 PSI provides a rudimentary way of navigating through a multi-program TS. Further metadata tables for EPG (electronic programme guides) and other applications have been defined in DVB SI (Digital Video Broadcasting Service Information) and ATSC PSIP (Advanced Television Systems Committee Program and System Information Protocol).

5.6 Current MPEG activities

In addition to the MPEG standards discussed in the previous sections, there are several more MPEG standards in various stages of the standardisation process. In MPEG-4 Part 10 (AVC), the SVC Profile has recently been added. With SVC, the same video signal is transmitted in two or more formats, for example QCIF and CIF. However, rather than coding the CIF signal on its own, it uses the QCIF signal as a base layer so that only the additional detail has to be coded. For more information on SVC, see Chapter 10.

In addition to video coding, there are several other standardisation activities within MPEG. MPEG-7 (ISO/IEC 15938), formally named 'Multimedia Content Description Interface', deals with content description and multimedia information retrieval. MPEG-21 (ISO/IEC 21000) is a 'Multimedia Framework Initiative'. Furthermore, a number of MPEG standards are being worked on, as shown in Table 5.9.

Table 5.9   Current MPEG activities

MPEG-A	Multimedia application formats
MPEG-B	MPEG system technologies
MPEG-C	MPEG video technologies
MPEG-D	MPEG audio technologies
MPEG-E	MPEG multimedia middleware

With the exception of MPEG-C, none of these standards deal with video compression algorithms and are therefore not relevant to this chapter. MPEG-C consists of three parts:

- *Part 1*: Accuracy specification for implementation of integer-output inverse DCT
- *Part 2*: Fixed-point implementation of DCT/inverse DCT
- *Part 3*: Auxiliary video data representation (stereoscopic video applications and requirements)

These are clarifications on the standards rather than additional video compression tools. However, this does not imply that there is currently no research activity on video compression. It just means that the algorithms being researched are not mature enough to be standardised.

ISO/IEC MPEG and ITU are coordinating their standardisation processes from time to time. Figure 5.16 shows that MPEG-2 is identical to ITU H.262, and MPEG-4 (AVC) is the same as H.264. On the other hand, H.261 has been developed in parallel with MPEG-1 and is similar but not identical to MPEG-1. Likewise, MPEG-4 (Visual) has similarities with H.263, but is not identical.

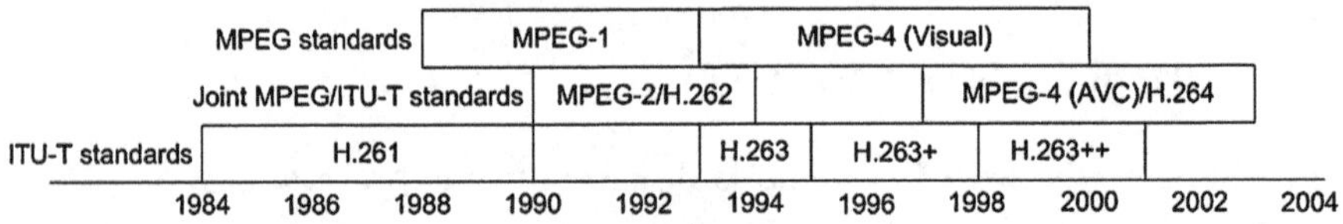

MPEG standards	MPEG-1		MPEG-4 (Visual)			
Joint MPEG/ITU-T standards		MPEG-2/H.262			MPEG-4 (AVC)/H.264	
ITU-T standards	H.261			H.263	H.263+	H.263++	

| 1984 | 1986 | 1988 | 1990 | 1992 | 1994 | 1996 | 1998 | 2000 | 2002 | 2004 |

Figure 5.16   Relationship between ITU and MPEG standards

Since MPEG-4 (AVC) and H.264 are identical, many publications now refer to H.264 instead of MPEG-4 (AVC). This is perfectly acceptable when no other coding standards are being considered. However, since we are dealing with several MPEG standards in this book, it is less confusing to use the MPEG nomenclature throughout.

5.7   Concluding remarks

This chapter presented a broad overview of all currently used MPEG video compression algorithms. In addition to the compression tools themselves, the chapter also gave a brief summary of bit-stream structures and the way audio and video bit streams are synchronised.

5.8   Summary

- MPEG specifies encoding tools, bit streams and how to decode bit streams, but it does not define encoding algorithms.

- The main differences between MPEG-1 and MPEG-2 are the interlace coding tools of MPEG-2.
- The Advanced Simple Profile of MPEG-4 (Visual) has been used for streaming applications, but MPEG-4 (Visual) is now all but superseded by MPEG-4 (AVC).
- MPEG-4 Part 10 (AVC), often referred to as H.264, should not be confused with MPEG-4 Part 2 (Visual).
- MPEG-2 Part 1 (Systems) is used for the transportation of MPEG-2 as well as MPEG-4 (AVC) bit streams.

Exercise 5.1   Encoder evaluation

A test engineer is evaluating two HDTV MPEG-4 (AVC) encoders. Comparing the data sheets (see Table 5.10), it is obvious that Encoder 1 has more coding tools than Encoder 2.

Table 5.10   Encoder compression tools

Coding tool	Encoder 1	Encoder 2
Motion estimation search range	1024×512	512×256
MBAFF	√	×
PAFF	√	√
Maximum number of B frames	4	3
CABAC entropy coding	√	√
CAVLC entropy coding	√	×
Spatial direct mode	√	√
Temporal direct mode	√	×
High profile	√	√
Scaling lists	Configurable	Fixed

Which encoder is likely to produce a better picture quality?

Exercise 5.2   Bit-rate saving

In order to reduce the bit rate of an MPEG-2 encoder, the horizontal resolution, i.e. the number of pixels per line, is often reduced to ¾. However, reducing the horizontal resolution to ¾ in MPEG-4 (AVC) encoders does not seem to have the same effect as in MPEG-2. Why does a reduction in horizontal resolution improve the picture quality of MPEG-2 more than it does in MPEG-4 (AVC)?

Exercise 5.3   PSNR limit

A test engineer measures the SNR of an MPEG-2 encoder–decoder system using a full-screen static test signal. As he/she increases the bit rate, the SNR rapidly approaches 50 dB but never exceeds 55 dB even at very high bit rates. What is the cause of this limitation?

References

1. International Standard ISO/IEC 11172-2 MPEG-1 Video, 1993.
2. D. LeGall, 'MPEG: A Video Compression Standard for Multimedia Applications', Communications of the ACM, April 1991. pp. 46–58.
3. P. Ikkurthy, M.A. Labrador, 'Characterization of MPEG-4 Traffic Over IEEE 802.11b Wireless LANs', IEEE Conference on Local Computer Networks, 2002. pp. 421–7.
4. International Standard ISO/IEC 13818-2 MPEG-2 Video, 1995.
5. Y. Senda, 'Approximate Criteria for the MPEG-2 Motion Estimation', *IEEE Transactions on Circuits and Systems for Video Technology*, Vol. 10, No. 3, April 2000. pp. 490–7.
6. International Standard ISO/IEC 14496-2 MPEG-4 Visual, 2004.
7. International Standard ISO/IEC 14496-10 MPEG-4 AVC, 2005.
8. D. Liu, D. Zhao, J. Sun, W. Gao, 'Direct Mode Coding for B Pictures Using Virtual Reference Picture', IEEE International Conference on Multimedia and Expo, Beijing, July 2007. pp. 1363–6.
9. Y. Su, M.-T. Sun, V. Hsu, 'Global Motion Estimation from Coarsely Sampled Motion Vector Field and the Applications', *IEEE Transactions on Circuits and Systems for Video Technology*, Vol. 15, Issue 2, February 2005. pp. 232–42.
10. Y. Keller, A. Averbuch, 'Fast Gradient Methods Based on Global Motion Estimation for Video Compression', *IEEE Transactions on Circuits and Systems for Video Technology*, Vol. 13, Issue 4, April 2003. pp. 300–9.
11. I.E.G. Richardson, *H.264 and MPEG-4 Video Compression, Video Coding for Next-Generation Multimedia*, Chichester: Wiley, ISBN 0470848375, 2003.
12. T. Muzaffar, T.S. Choi, 'Maximum Video Compression Using AC-Coefficient Prediction', Proceedings of the IEEE Region 10 Conference, South Korea, 1999. pp. 581–4.
13. M. van der Schaar, *MPEG-4 Beyond Conventional Video Coding: Object Coding, Resilience and Scalability*, San Rafael, CA: Morgan & Claypool Publishers, ISBN 1598290428, 2006.
14. IEEE Std 1180–1990, 'IEEE Standard Specification for the Implementations of 8×8 Inverse Discrete Cosine Transform', Institute of Electrical and Electronics Engineers, New York, USA, International Standard, December 1990.
15. C.-P. Fan, 'Fast 2-Dimensional 4×4 Forward Integer Transform Implementation for H.264/AVC', *IEEE Transactions on Circuits and Systems*, Vol. 53, Issue 3, March 2006.

16. D. Marpe, S. Gordon, T. Wiegand, 'H.264/MPEG-4 AVC Fidelity Range Extensions: Tools, Profiles, Performance, and Application Areas', IEEE International Conference on Image Processing, September 2005. pp. I/593–6.

17. P.G. Howard, J.S. Vitter, 'Arithmetic Coding for Data Compression', *Proceedings of the IEEE*, Vol. 82, Issue 6, June 1994. pp. 749–64.

18. D. Marpe, H. Schwarz, T. Wiegand, 'Context-Based Adaptive Binary Arithmetic Coding in the H.264/AVC Video Compression Standard', *IEEE Transactions on Circuits and Systems for Video Technology*, Vol. 13, Issue 7, July 2003. pp. 620–36.

19. M. Liu, 'Multiple Description Video Coding Using Hierarchical B Pictures', IEEE International Conference on Multimedia and Expo, Beijing, July 2007. pp. 1367–70.

20. N. Ozbek, A.M. Tekalp, 'H.264 Long-Term Reference Selection for Videos with Frequent Camera Transitions', *IEEE Signal Processing and Communications Applications*, April 2006. pp. 1–4.

21. J.M. Boyce, 'Weighted Prediction in the H.264/MPEG AVC Video Coding Standard', *Proceedings of the International Symposium on Circuits and Systems*, Vol. 3, May 2004. pp. III/789–92.

22. A. Major, I. Nousias, S. Khawam, M. Milward, Y. Yi, T. Arslan, 'H.264/AVC In-Loop De-Blocking Filter Targeting a Dynamically Reconfigurable Instruction Cell Based Architecture', Second NASA/ESA Conference on Adaptive Hardware and Systems, Edinburgh, August 2007. pp. 134–8.

23. G.J. Sullivan, T. Wiegand, 'Rate-Distortion Optimization for Video Compression', *IEEE Signal Processing Magazine*, Vol. 15, Issue 6, November 1998. pp. 74–90.

24. J. Zhang, X. Yi, N. Ling, W. Shang, 'Chroma Coding Efficiency Improvement with Context Adaptive Lagrange Multiplier (CALM)', IEEE International Symposium on Circuits and Systems, May 2007. pp. 293–6.

Chapter 6

Non-MPEG compression algorithms

6.1 Introduction

Despite the fact that MPEG and the closely related ITU H.26x video compression standards are widely used in broadcast, telecommunications, IPTV as well as many other applications, there are, of course, numerous other video compression algorithms, each with its own advantages and niche applications. Although it goes beyond the scope of this book to introduce all commonly used compression algorithms, it is worth having a look at a few of the more important ones in order to show different approaches and algorithm variations.

In particular, the Video Codec 1 (VC-1) algorithm, developed by Microsoft, is a good example of a vendor-driven, block-based, motion-compensated compression algorithm with some interesting differences to MPEG algorithms. Since the decoding algorithm has been standardised in Society of Motion Picture and Television Engineers (SMPTE), most of its compression tools are now in the public domain. Suitable for coding interlaced video signals and fully integrated with other Microsoft products, it provides an alternative for IPTV streaming applications.

A second algorithm worth mentioning is the Chinese Audio Video Coding Standard (AVS). Although it has many similarities with MPEG-4 (AVC), it avoids some of the most processing-intensive parts of the MPEG-4 (AVC) standard.

In contrast to the block-based VC-1 and AVS algorithms, which have many similarities with MPEG algorithms, it is worth also looking at some non-block-based compression algorithms, i.e. algorithms based on wavelet technology. Wavelet transforms are quite different from block-based transforms. Therefore, a brief summary about wavelet theory is provided in Appendix E. Two algorithms need to be examined in more detail: JPEG2000, which is becoming increasingly more relevant to broadcast applications, and Dirac, a family of open-source algorithms developed primarily by the BBC research department.

6.2 VC-1 SMPTE 421M

VC-1 is the first video compression algorithm standardised in SMPTE. It is a bit stream and decoding definition for the compression algorithm originally known as the Microsoft Windows Media 9. Windows Media^{TM} Codecs have been developed in parallel with the MPEG and ITU standards. Whereas WMV-7 and WMV-8 had been developed for small, progressive IPTV streaming applications, WMV-9 (Advanced Profile) is capable of coding interlaced video signals and is, therefore, suitable for broadcast applications. Note that WMV-9 is the Microsoft encoder algorithm, compatible with VC-1 decoding.

Because it has been optimised for real-time decoding on state-of-the-art PCs, VC-1 does not support the same coding tool set as MPEG-4 (AVC) although there are many similarities with both MPEG-4 (Visual) and MPEG-4 (AVC). For example, VC-1 uses multiple VLC code tables instead of CAVLC or CABAC. Nevertheless, its compression performance, particularly on film material, is very impressive.

Some of the main differences between VC-1 and MPEG-4 (AVC) are explained in Section 6.2.1. For a more detailed description of the Windows Media Video 9 algorithm, see Reference 1.

6.2.1   Adaptive block-size transform

Whereas MPEG-4 (AVC) High Profile defines two distinct integer transforms with block sizes of 4×4 and 8×8, VC-1 uses an adaptive integer transform that supports 8×8, 8×4, 4×8 and 4×4 block sizes. The inverse transforms can be carried out in 16 bit arithmetic which helps to speed up decoding. Figures 6.1 and 6.2 show the transform kernels for the 8×8 and 4×4 inverse transformations, respectively.

```
12   12   12   12   12   12   12   12
16   15    9    4   -4   -9  -15  -16
16    6   -6  -16  -16   -6    6   16
15   -4  -16   -9    9   16    4  -15
12  -12  -12   12   12  -12  -12   12
 9  -16    4   15  -15   -4   16   -9
 6  -16   16   -6   -6   16  -16    6
 4   -9   15  -16   16  -15    9   -4
```

Figure 6.1 8×8 kernel for VC-1 inverse transformation

```
17   17   17   17
22   10  -10  -22
17  -17  -17   17
10  -22   22  -10
```

Figure 6.2 4×4 kernel for VC-1 inverse transformation

6.2.2   Motion compensation

The smallest motion compensation block size in VC-1 is deliberately limited to 8×8 blocks in order to reduce the processing requirement on decoders. A motion compensation block size of 4×4 pixels needs four times as many memory access cycles as an 8×8 block.

6.2.3 Advanced entropy coding

Instead of CAVLC or CABAC entropy coding, VC-1 uses multiple VLC code tables for the encoding of transform coefficients.

6.2.4 De-blocking filter

VC-1 has two methods for reducing blocking artefacts in critical sequences: it has an in-loop filter that operates on all picture types and an overlap-smoothing algorithm specifically designed for intra-coded block edges.

6.2.5 Advanced B frame coding

B frame coding in VC-1 is similar to MPEG-4 (Visual), except that VC-1 allows bottom-field B pictures to take predictions from the top-field picture of the same B frame.

6.2.6 Low-rate tools

VC-1 can operate at extremely low bit rates. As will be shown in Chapter 12, compressed video signals need high bit rates for short periods of time. Most of the time, a lower bit rate would be sufficient. When VC-1 is set to extremely low bit rates, e.g. below 100 kbit/s, it can downsize critical scenes on a frame-by-frame basis. Although this reduces spatial resolution, it avoids blocking artefacts.

6.2.7 Fading compensation

Fades to or from black are often difficult to code because the changes in brightness can confuse the motion estimation algorithm. To eliminate this problem VC-1 uses a fading compensation algorithm, which adapts the brightness level for the purpose of motion compensation. The fading parameters are sent to the decoder to restore the original brightness level.

6.3 Audio Video Coding Standard (AVS)

The AVS was developed independently of MPEG. Although there are some similarities with MPEG-4 (AVC), there are also many differences. AVS-P2 (Part 2 video) is a highly efficient second-generation video coding standard but is in many ways somewhat simpler and easier to implement than MPEG-4 (AVC). Some of the important differences between AVS and MPEG-4 (AVC) are explained in Section 6.3.1. For a more detailed description of the AVS video coding algorithm, see Reference 2.

6.3.1   Transform

AVS uses a single 8 × 8 integer transform, which is different from the High Profile 8 × 8 transform of MPEG-4 (AVC). Figure 6.3 shows the transform kernel for AVS.

```
8   10   10    9    8    6    4    2
8    9    4   -2   -8  -10  -10   -6
8    6   -4  -10   -8    2   10    9
8    2  -10   -6    8    9   -4  -10
8   -2  -10    6    8   -9   -4   10
8   -6   -4   10   -8   -2   10   -9
8   -9    4    2   -8   10  -10    6
8  -10   10   -9    8   -6    4   -2
```

Figure 6.3   8 × 8 kernel used in AVS transformation

6.3.2   Intra-predictions

Instead of the nine intra-prediction modes defined in MPEG-4 (AVC), AVS supports the five most important 8 × 8 intra-prediction modes in luma and 4 chroma modes. The luma modes are horizontal, vertical, down left, down right and DC. The chroma modes are horizontal, vertical, plane and DC. Both luma and chroma prediction modes include a low-pass filter to reduce noise predictions.

6.3.3   Motion compensation

AVS supports motion compensation block sizes of 16 × 16, 16 × 8, 8 × 16 and 8 × 8.

6.3.4   Entropy coding

AVS uses adaptive VLC coding with 19 different VLC tables.

6.3.5   De-blocking filter

Since AVS uses only 8 × 8 transforms, this reduces the complexity of the de-blocking filter considerably because there are fewer block boundaries to be filtered.

6.3.6   B frame coding modes

In addition to an improved direct temporal mode, AVS supports a 'symmetric mode', which allows bi-directional predictions with a single motion vector pair [3].

Table 6.1 gives a summarised comparison of the coding tools used in AVS and VC-1 in relation to all four MPEG video compression standards. Comparisons between different video codecs have been carried out with thorough analysis of processing power and compression efficiency [4–6]. However, one should not read too much into such evaluations because the performance of codecs improves over time [7] and any such comparison can only represent a snapshot in time.

Non-MPEG compression algorithms 89

Table 6.1 Comparison of coding tools between MPEG-1, MPEG-2 Main Profile, MPEG-4 (Visual) Main Profile, MPEG-4 (AVC) High Profile, VC-1 Advanced Profile and AVS

Coding tool	MPEG-1	MPEG-2 Main Profile	MPEG-4 (Visual) Main Profile	MPEG-4 (AVC) High Profile	VC-1 (Advanced Profile)	AVS
Transform	8×8 DCT	8×8 DCT	8×8 DCT	4×4 or 8×8 integer transform	$8 \times 8, 8 \times 4, 4 \times 8, 4 \times 4$ integer DCT	8×8 integer DCT
Intra prediction	DC only	DC only	AC prediction in frequency domain	$16 \times 16, 8 \times 8, 4 \times 4$ spatial predictions	AC prediction in frequency domain	8×8 spatial predictions
Interlace coding tools	No	Yes	Yes	Yes	Yes	Yes
Motion compensation block sizes	16×16	$16 \times 16, 16 \times 8$	$16 \times 16, 16 \times 8, 8 \times 16, 8 \times 8$	$16 \times 16, 16 \times 8, 8 \times 16, 8 \times 8, 8 \times 4, 4 \times 8, 4 \times 4$	$16 \times 16, 16 \times 8, 8 \times 16, 8 \times 8$	$16 \times 16, 16 \times 8, 8 \times 16, 8 \times 8$
Motion compensation beyond picture	No	No	Yes	Yes	Yes	Yes
Motion vector precision	½ pel	½ pel	¼ pel	¼ pel	¼ pel	¼ pel
Multiple reference P frames	No	No	No	Yes	No	Yes
Weighted prediction	No	No	No	Yes	Fading compensation	No

Table 6.1 Continued

Coding tool	MPEG-1	MPEG-2 Main Profile	MPEG-4 (Visual) Main Profile	MPEG-4 (AVC) High Profile	VC-1 (Advanced Profile)	AVS
Direct mode predictions on B frames	No	No	Yes	Spatial and temporal	Yes	Temporal and symmetric
Predictions from B frames	No	No	No	Yes	Yes	No
De-blocking filter	No	No	Optional post-processing filter	Mandatory in-loop filter	In-loop filter and overlap smoothing	Mandatory in-loop filter
Entropy coding	Fixed VLC	Fixed VLC	Fixed VLC	CABAC	Adaptive VLC	Adaptive VLC

6.4 Wavelet compression

Originally developed for speech processing [8], wavelets are increasingly finding their way into image-processing applications such as texture analysis [9] and synthesis [10], as well as image compression algorithms. Similar to sub-band coding and other signal-processing techniques, wavelet transformation was originally developed to overcome certain difficulties in the areas of acoustic analysis and speech synthesis [11]. Since then, wavelets have been applied in such diverse areas as non-destructive measurement [12], laser interferometry [13], pattern recognition [14], robotic vision [15] and, of course, video compression [16]. More recently, the performance of wavelet coding has been significantly improved [17,18].

Appendix E gives a brief introduction of wavelet theory. Instead of using DCT-based transformation on a block-by-block basis, wavelet transformation is applied to the entire image as a whole, both in horizontal and vertical direction. Unlike Fourier transformation that produces a two-dimensional spectrum of complex coefficients, wavelet transformation decomposes the input signal into low- and high-frequency bands.

In most image-processing applications, as well as in image compression, wavelet filters are applied consecutively in horizontal and vertical direction in order to decompose the image into four sub-images: a low-low image, which has been low-pass filtered in horizontal and vertical direction; a low-high image, which contains low horizontal but high vertical frequencies; a high-low image, which contains high horizontal but low vertical frequencies; and a high-high image, which contains high horizontal and vertical frequencies. Figure 6.4 shows a block

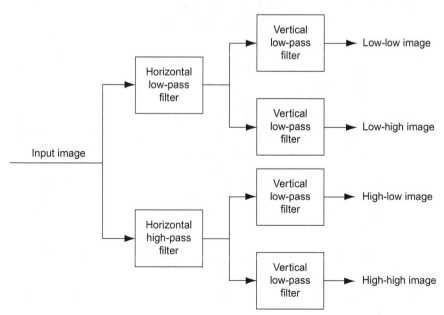

Figure 6.4   Block diagram of a single-stage two-dimensional decomposition filter

diagram of a single-stage decomposition filter. This process of decomposition can be repeated several times by taking the low-low image as the input of the next stage. In the decoder, the corresponding reconstruction filters are used to regenerate the original image.

Note that with each stage of two-dimensional decomposition, the bit accuracy of the down-sampled images have to be increased by a factor of 4 in order to achieve perfect reconstruction of the original image. Therefore, if three decomposition stages are used on an 8 bit image, the required bit accuracy in the low-low image increases by a factor of $4 \times 4 \times 4 = 64$, i.e. from 8 to 14 bits. As with DCT, transformation itself increases bit-rate requirements. The savings are achieved through quantisation. In general, the high-high images can be quantised more heavily than the low-low images.

A second important observation is that the low-low image is simply a down-sized image of the original. For that reason, spatial scalability is inherently part of intra-coding wavelet compression engines.

6.5   JPEG2000

JPEG2000 is the latest series of standards from the Joint Photographic Experts Group (JPEG) committee. Its compression efficiency is about 20 per cent better than that of the original JPEG standard developed in 1991. JPEG2000 is becoming increasingly more important for high-quality broadcast-contribution applications because it supports 10 bit accuracy and spatial scalability [19].

JPEG2000 is a picture coding algorithm based on wavelet compression and arithmetic coding. Therefore, its compression artefacts are quite different from those of MPEG-1 and MPEG-2. Instead of generating blocking artefacts, JPEG2000 reduces resolution if there are not enough bits to encode the full detail. This is done by quantising coefficients in the high-frequency bands more heavily than those in the low-frequency bands [20]. In a way its compression artefacts are more akin to those of MPEG-4 (AVC), assuming that the in-loop filter is enabled.

Used for video compression, it is often referred to as Motion JPEG2000 [21], since it is a still image compression algorithm and does not use motion prediction. As a result, JPEG2000 generally requires significantly higher bit rates than MPEG-4 (AVC) compression. However, the relative bit-rate demand on individual sequences is very different from MPEG-4 (AVC). What can be classed as a non-critical sequence in MPEG-4 (AVC) could be a highly critical sequence for JPEG2000 and vice versa. For example, the bit-rate demand of MPEG-4 (AVC) for a sequence with high detail but little or smooth motion is relatively low whereas JPEG2000 would require a relatively high bit rate for such a sequence. Conversely, a sequence with little detail but complex motion requires a relatively high bit rate in MPEG-4 (AVC) and a relatively low bit rate in JPEG2000.

These differences are highlighted in Figures 6.5 and 6.6. Figure 6.5 shows the rate-distortion curves for the 'Mobile and Calendar' sequence, which contains significant amounts of detail but has slow motion. On the other hand, the 'Soccer'

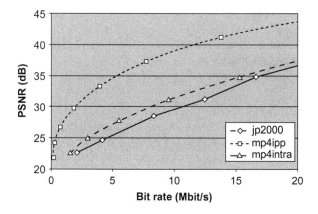

*Figure 6.5 PSNR comparison between JPEG2000 and MPEG-4 (AVC) on a
sequence with slow motion (Mobile and Calendar); mp4intra:
MPEG-4 (AVC) intra-coding only; mp4ipp: MPEG-4 (AVC)
predictive coding (without B frames)*

sequence shown in Figure 6.6 has less detail but complex motion. As a first
approximation, the performance of JPEG2000 is quite similar to that of MPEG-4
(AVC) intra-coding, i.e. not using P or B frames. Figures 6.5 and 6.6 show that
JPEG2000 uses approximately 10–20 per cent more bit rate than MPEG-4 (AVC)
intra-coding. Similar results have been published in a detailed study on JPEG2000
and MPEG-4 (AVC) bit-rate efficiency [22].

Once motion-compensated P frames are enabled in MPEG-4 (AVC), the situation becomes quite different. Now we find that JPEG2000 needs 450 per cent

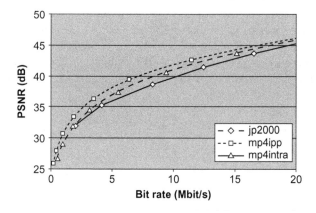

*Figure 6.6 PSNR comparison between JPEG2000 and MPEG-4 (AVC) on a
sequence with complex motion (Soccer); mp4intra: MPEG-4 (AVC)
intra-coding only; mp4ipp: MPEG-4 (AVC) predictive coding
(without B frames)*

more bit rate for 'Mobile and Calendar' but 'only' 66 per cent more for 'Soccer'. Note that in this comparison, B frames have not been enabled in MPEG-4 (AVC) in order to keep the end-to-end delay low. By using B frames, the bit-rate demand of MPEG-4 (AVC) could be reduced further.

PSNR comparisons between algorithms which produce different types of artefacts are, of course, highly problematic and one should not read too much into these comparisons (see Chapter 3). The main point to make is that JPEG2000 is roughly equivalent to MPEG-4 (AVC) intra-compression but MPEG-4 (AVC) with P and B frames can achieve much lower bit rates for the same video quality.

6.6   Dirac

Dirac is a family of wavelet-based, open-source video compression algorithms developed primarily by the BBC research department [23]. The intention was to develop a series of royalty-free video codecs for professional applications. There are, of course, countless numbers of patents related to video compression, and it is almost impossible to develop a compression algorithm that does not infringe some of those patents. The problem was solved in MPEG by creating a patent pool for each of the MPEG algorithms. The patent pools are administered by the MPEG Licence Authority (MPEG LA).

The original Dirac algorithm was aiming for highest compression efficiency, whereas later versions of the algorithm, referred to as 'Dirac Pro', were aiming at production and post-production applications with low end-to-end delays [24]. The most efficient version of the algorithm in terms of video compression uses block-based motion compensation, wavelet transformation and arithmetic coding. Figure 6.7 shows a block diagram of this version of a Dirac encoder.

There is a potential problem in combining block-based motion compensation with wavelet transforms. The difficulty is that arbitrarily shaped moving objects produce an error signal containing high-frequency block edges in block-based motion-compensated predicted pictures. This is because natural-shaped objects cannot be properly motion compensated using rectangular blocks. The same is true in MPEG compression systems, but the difference is that in MPEG both motion

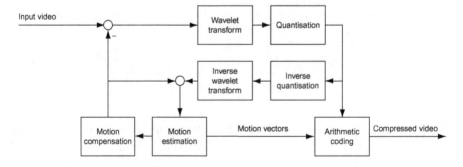

Figure 6.7   Block diagram of a Dirac encoder with motion compensation

compensation and transformation are block based. Therefore, motion compensation and transformation are block aligned so that most block edges are not part of transformation.

In the Dirac encoder, however, the motion-compensated error signal containing the high-frequency block edges are compressed using wavelet transforms. High-frequency block edges manifest themselves as high-level content in the high-frequency bands and are therefore difficult to compress using wavelet transforms. To ameliorate this problem, Dirac uses overlapped block-based motion compensation, which reduces the high-frequency content of the block edges.

A second fundamental problem with using wavelet transformation for video compression is to do with interlace. While block-based transformations offer a choice between field or frame predictions and field or frame transformations on a macroblock-by-macroblock basis, wavelet codecs have to transform an entire video frame either as one frame picture or as two field pictures. Most video material, however, has a mix of relatively static backgrounds and more or less fast-moving objects.

6.6.1 Dirac Pro

Dirac Pro are simplified versions of the original Dirac algorithm. There are currently two versions of Dirac Pro: Dirac Pro 1.5 and Dirac Pro 270. Both are being standardised through SMPTE as Video Codec 2 (VC-2) and made available as open-technology video codecs. Still based on wavelet transforms, these are simpler versions of the above algorithm, in that they do not use motion compensation but have much lower latency. Dirac Pro 1.5 is intended to compress a 1080p50/60 video signal down to 1.5 Gbit/s such that it can be transported over a conventional HD-SDI link. Dirac Pro 270 achieves a higher compression ratio in order to carry a 1080p50/60 video signal in a 270 Mbit/s SD–SDI link. Whereas Dirac Pro 1.5 uses a simple two-level Haar wavelet transform and exp-Golomb variable length coding, Dirac Pro 270 uses a three-level LeGall wavelet transform and arithmetic coding in order to achieve the higher compression ratio. Figure 6.8 shows a generic block diagram of a Dirac Pro encoder.

6.7 Concluding remarks

In addition to the examples illustrated above, there are, of course, countless other codecs for video compression, many of them using proprietary algorithms or variations on standardised techniques. Since these algorithms are not in the public domain, proper comparisons cannot be carried out. Nevertheless, the coding tools explained in the previous chapter on MPEG compression, together with the non-

Figure 6.8 Generic block diagram of a Dirac Pro encoder

MPEG algorithms explained in this chapter should provide a broad overview of currently used video compression techniques.

6.8 Summary

- WMV-9 is the Microsoft encoder algorithm, compatible with VC-1 decoding.
- AVS has many similarities with MPEG-4 (AVC), but it avoids some of the most processing-intensive algorithms.
- Wavelet transformation is not as compatible to block-based motion compensation and interlaced video as block-based transformation.
- Sequences with fast, complex motion, which may be difficult to code with motion-compensated algorithms, are not necessarily critical for JPEG2000.
- Sequences with high detail and slow motion are critical for JPEG2000 but not for motion-compensated codecs.

Exercise 6.1   Clip encoding

Which coding standards are suitable for long-GOP QVGA clip encoding and why?

Exercise 6.2   Codec comparison

A telecommunications company is evaluating the compression algorithms listed in this chapter and comparing them to an MPEG-4 (AVC) encoder. Could this evaluation be carried out with PSNR measurements?

References

1. S. Srinivasan, P. Hsu, T. Holcomb, K. Mukerjee, L. Shankar, 'Windows Media Video 9: Overview and Applications', *Signal Processing: Image Communication*, Vol. 19, Issue 9, October 2004. pp. 851–75.
2. L. Fan, S. Ma, F. Wu, 'Overview of AVS Video Standard', IEEE International Conference on Multimedia and Expo, Taipei, June 2004. pp. 423–6.
3. X. Ji, D. Zhao, F. Wu, Y. Lu, W. Gao, 'B-Picture Coding in AVS Video Compression Standard', *Signal Processing: Image Communication*, Vol. 23, Issue 1, January 2008. pp. 31–41.
4. J. Bennett, A.M. Bock, 'In-Depth Review of Advanced Coding Technologies for Low Bit Rate Broadcast Applications', *SMPTE Motion Imaging Journal*, Vol. 113, December 2004. pp. 413–8.
5. P. Lambert, W. DeNeve, P. DeNeve, I. Moerman, P. Demeester, R. Van de Walle, 'Rate-Distortion Performance of H.264/AVC Compared to State-of-the-art Video Codecs', *IEEE Transactions on Circuits and Systems for Video Technology*, Vol. 16, Issue 1, January 2006. pp. 134–40.
6. M.A. Isnardi, 'Historical Overview of Video Compression in Consumer Electronic Devices', International Conference on Consumer Electronics, January 2007.

7. M. Goldman, A.M. Bock, 'Advanced Compression Technologies for High Definition', IEE Conference on 'IT to HD', December 2005.

8. R.P. Ramachandran, R. Mammone, *Modern Methods of Speech Processing*, New York: Kluwer Academic Publishers, ISBN 0792396073, 1995.

9. S. Liu, H. Yao, W. Gao, 'Steganalysis Based on Wavelet Texture Analysis and Neural Network', 5th World Congress on Intelligent Control and Automation, June 2004. pp. 4066–9.

10. L. Tonietto, M. Walter, C.R. Jung, 'Patch-Based Texture Synthesis Using Wavelets', 18th Brazilian Symposium on Computer Graphics and Image Processing, October 2005. pp. 383–9.

11. M. Kobayashi, M. Sakamoto, T. Saito, Y. Hashimoto, M. Nishimura, K. Suzuki, 'Wavelet Analysis Used in Text-to-Speech Synthesis', *IEEE Transactions on Circuits and Systems*, Vol. 45, Issue 8, August 1998. pp. 1125–9.

12. Y. Shao, Y. He, 'Nondestructive Measurement of the Acidity of Strawberry Based on Wavelet Transform and Partial Least Squares', Conference Proceedings on Instrumentation and Measurement Technology, Warsaw, May 2007.

13. L.R. Watkins, S.M. Tan, T.H. Barnes, 'Interferometer Profile Extraction Using Continuous Wavelettransform', *Electronics Letters*, Vol. 33, Issue 25, December 1997. pp. 2116–7.

14. G.Y. Chen, W.F. Xie, 'Pattern recognition Using Dual-Tree Complex Wavelet Features and SVM', Canadian Conference on Electrical and Computer Engineering, May 2005. pp. 2053–6.

15. I.E. Rube, M. Ahmed, M. Kamel, 'Affine Invariant Multiscale Wavelet-Based Shape Matching Algorithm', 1st Canadian Conference on Computer and Robot Vision, May 2004. pp. 217–24.

16. D. Lazar, A. Averbuch, 'Wavelet Video Compression Using Region Based Motion Estimation and Compensation', IEEE International Conference on Acoustics, Speech, and Signal Processing, 2001. pp. 1597–600.

17. J.-R. Ohm, 'Advances in Scalable Video Coding', *Proceedings of IEEE*, Vol. 94, Issue 1, January 2005. pp. 42–56.

18. D. Wang, L. Zhang, R. Klepko, A. Vincent, 'A Wavelet-Based Video Codec and Its Performance', International Broadcasting Convention, Amsterdam, September 2007. pp. 340–7.

19. D. Taubman, M. Marcellin, *JPEG2000: Image Compression Fundamentals, Standards and Practice*, New York: Kluwer Academic Publishers, ISBN 079237519X, 2001.

20. M. Long, H.-M. Tai, S. Yang, 'Quantisation Step Selection Schemes in JPEG2000', *Electronics Letters*, Vol. 38, Issue 12, June 2002. pp. 547–9.

21. T. Fukuhara, K. Katoh, S. Kimura, K. Hosaka, A. Leung, 'Motion-JPEG2000 Standardization and Target Market', International Conference on Image Processing, Vancouver, September 2000. pp. 57–60.

22. M. Ouaret, F. Dufaux, T. Ebrahimi, 'On Comparing JPEG2000 and Intraframe AVC', *Proceedings of the SPIE*, September 2006.

23. T. Borer, T. Davies, A. Suraparaju, 'Dirac Video Compression', International Broadcasting Convention, Amsterdam, September 2005. pp. 201–8.
24. T. Borer, 'Open Technology Video Compression for Production and Post Production', International Broadcasting Convention, Amsterdam, September 2007. pp. 324–31.

Chapter 7

Motion estimation

7.1 Introduction

Motion estimation is arguably one the most important sub-functions of any motion-compensated video compression algorithm – a reason enough to allocate a whole chapter to it. Although there are differences in terms of prediction modes and block sizes, the removal of temporal redundancy in video signals inevitably requires a search engine that provides motion information on predicted blocks or block partitions. However, motion estimation techniques are not only a major part of video compression algorithms, they are also used in noise reduction [1], de-interlacing [2] (see also Chapter 8), standards conversion [3] and many other video-processing applications. It is, therefore, not surprising that a large number of motion estimation algorithms have been developed for video compression as well as for other application areas [4].

In terms of video objects, we can distinguish several types of motion:

- *Pan*: The camera moves in one direction, often tracking the movement of an object, for example a walking person. This results in translational motion of the background even though the object itself is relatively stationary. Although translational motion can be readily compensated, it is often the sheer speed of motion that can cause problems in fast camera pans.
- *Zoom*: The camera homes in on an object or zooms out to give a wider angle of view. This results in non-translational motion, whereby objects grow or shrink in size. Since most motion-compensated compression algorithms only support translational motion, zooms tend to generate significant prediction errors.
- *Rotational motion*: The camera captures an object that rotates. Rotational motion also produces significant prediction errors in translational motion-compensated systems.
- *Random motion*: This is the most difficult type of motion because it contains the least amount of temporal redundancy. Typical examples are splashing water or noise-like effects.

Given that the majority of types of motion seem to be non-translational, why do compression algorithms only support translational motion compensation? Or, to put it another way, how can compression algorithms remove temporal redundancy so effectively, if they cannot track true motion?

The answer to the first question is simple and was already hinted at in Chapter 4. Even translational motion estimation can be extremely processing intensive, if large search ranges are to be covered using relatively simple block-matching algorithms. To search for all possibilities of affine motion is simply beyond the capabilities of today's real-time search engines.

The answer to the second question lies in the fact that motion estimation algorithms for video compression systems need not find true motion; they only need to find relatively good predictions for each block, no matter where the prediction comes from. Apart from that, smaller block sizes can help to improve predictions in areas of complex motion. Even if the prediction is not perfect, the prediction error signal can restore the reconstructed image to an acceptable quality.

Therefore, motion estimation for compression is somewhat simpler than motion estimation for noise reduction, de-interlacing and standards conversion algorithms. In those applications, motion vectors are used to move objects into new spatial and/or temporal positions, for which there is no reference in the video source. Consequently, there is no 'error correcting signal', which helps to hide prediction errors. Thus, wrong predictions result in visible artefacts. Nevertheless, the same motion estimation algorithms are often used for video compression as well as for other video-processing applications.

There are two fundamentally different types of motion estimation algorithms: block-matching algorithms in the spatial domain and correlation algorithms in the two-dimensional (spatial) frequency domain. Block-matching algorithms are mathematically simpler, but require high processing power to achieve a large search range. Correlation algorithms can achieve higher (sub-pixel) accuracy and can provide multiple motion vector candidates in one calculation. However, they do require two-dimensional Fourier transform operations, which are also quite processing intensive.

7.2   Block-matching algorithms in the spatial domain

7.2.1   Exhaustive motion estimation

The simplest, and also the most processing-intensive, block-matching algorithm is one that is exhaustive or enables full search over the entire search range. Figure 7.1 shows the frame order and the motion estimation from a reference frame to a coded P frame and Figure 7.2 illustrates the exhaustive search algorithm. Using the sum of absolute differences (SAD) between the predicted block and the search block as a match criterion, exhaustive motion estimation examines all possible search positions within a pre-defined search area and uses the one with the lowest SAD (best match) as the prediction candidate. Similarly, Figure 7.3 illustrates the motion estimation process on a B frame.

Since the processing power requirements of exhaustive motion estimation increases with the square of the search range, the question arises as to how big the

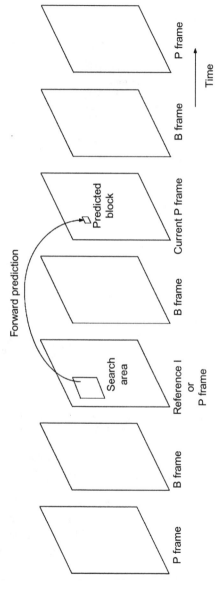

Figure 7.1 Motion estimation from a reference frame to the current P frame

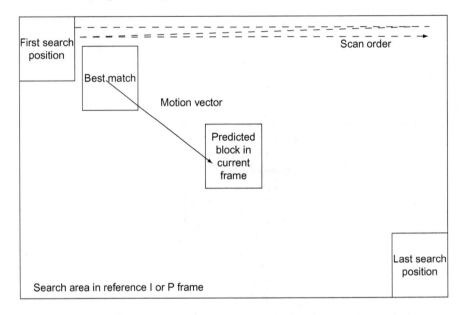

Figure 7.2   Exhaustive search algorithm

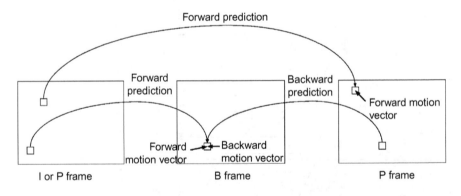

Figure 7.3   Motion estimation on a B frame

search range needs to be. Examining fast sports material on a frame-by-frame basis, it can easily be seen that motion displacements of 100 pixels per frame (in SDTV) are not uncommon. A more comprehensive answer can be found by examining a large number of test sequences and calculating the statistics across all motion vectors.

Figure 7.4 shows the logarithm of the probability density function of motion vectors up to a search range of 200 pixels. It can be seen that up to a search range of about 100 pixels per frame, the probability density drops off more or less monotonically. These vectors are derived from real moving objects. Above that value,

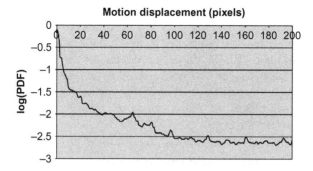

Figure 7.4   Motion vector statistics in SDTV

the probability density stays roughly the same, i.e. motion vectors of 200 pixels are almost similar to those of 100 pixels. This means that vectors above 100 pixels are more or less random hits, not necessarily related to real motion.

Figure 7.5 shows the computational cost of exhaustive motion estimation, normalised to the computational cost of the entire MPEG-2 encoding process, as a function of search range. It can be seen that for a search range of ±100 pixels, the computational cost of motion estimation is 500 times that of the rest of MPEG-2 encoding. As a practical example, we can assume a search range of ±100 pixels horizontally and ±64 lines vertically from one picture to the next. In this case we need to calculate $201 \times 129 \times 16 \times 16 = 6\,637\,824$ absolute difference values for each 16×16 macroblock (in MPEG-2 motion vectors apply to macroblocks of 16×16 luma samples). In 625-line standard definition, each video frame consists of $45 \times 36 = 1\,620$ macroblocks, so that we need to carry out a total of $102 \times 129 \times 16 \times 16 \times 1\,620 \times 23 = 247\,325\,322\,240$ absolute difference-accumulation calculations per second (assuming 2 intra-coded frames/s).

Apart from the enormous processing power requirement, unconstrained exhaustive motion estimation has the disadvantage that it tends to produce chaotic motion vectors in highly critical scenes or in low detailed areas with some source noise. Finding the best prediction based on the lowest SAD does not guarantee the

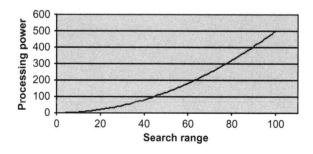

Figure 7.5   Computational cost of exhaustive motion estimation

best coding performance. This is because there is a danger that more bits are spent on motion vectors than are being saved on DCT coefficients. However, the problem of chaotic motion vectors can be avoided by using additional search criteria or rate-distortion optimisation (RDO) [5,6]. Nevertheless, it is the processing power requirement of exhaustive motion estimation that remains the biggest problem. Therefore, the majority of research effort in block-matching algorithms is devoted to trying to reduce processing power requirements without degrading the performance [7,8].

7.2.2   Hierarchical motion estimation

One of the most popular methods for reducing the processing requirements of block-matching algorithms is to downsize the image several times to produce a hierarchical structure of images [9,10]. Down-sampling reduces computational requirements for two reasons. Not only are the block sizes smaller, cutting down on the number of calculations necessary for each search position, but pixel displacements in the smaller images also correspond to larger displacements in the base layer and, therefore, a small search range in the smallest image corresponds to a larger search range in the base layer. For typical SDTV search ranges, the saving in processing power can be up to a factor of 500!

Figure 7.6 shows a block diagram of a hierarchical motion estimation system. Note that the 1:2 down-sample filter blocks are re-sizing both the reference image (to be coded) and the (previously coded) search image in both horizontal and vertical directions. Each down-sample block therefore represents four one-dimensional down-sampling filters. The motion estimation process starts off at the smallest image (quarter-quarter common intermediate format (QQCIF) in this case) to produce candidate vectors MV1 for next larger image. At each stage, the vectors found at the higher layer are used as a starting point for a refinement search at the current layer, whereby the refinement search range is much smaller than the total search range, typically only a small number of pixel positions in each direction. At the bottom layer motion vectors MV4 are the pixel-accurate vectors used for motion compensation.

While hierarchical motion estimation can give good performance in many applications where processing power is at a premium, there are a number of

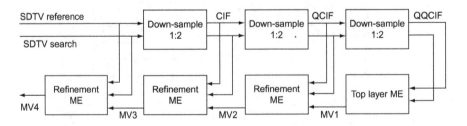

Figure 7.6   Block diagram of a hierarchical motion estimation system

disadvantages and potential problems with this form of motion estimation. The most obvious deficiency arises from the fact that the search in the higher layers can get trapped in local minima and it is often difficult to recover from sub-optimum starting points at the lower layers. Local minima occur in areas of periodic patterns, e.g. windows on a building. Figure 7.8a shows an example of an image with periodic patterns that can confuse hierarchical search algorithms.

A second disadvantage in a real-time implementation of hierarchical motion estimation is the fact that the higher layers of the motion estimation process have to be carried out on the source picture because in a real-time system there is not enough time to do a search on all layers of the reconstructed picture, i.e. within the coding loop. In high bit-rate applications, this does not usually cause a problem because the reconstructed video will be very similar to the source. If encoders with hierarchical motion estimation are pushed to extremely low bit rates, however, any discrepancy between source and reconstructed video will reduce the performance of the motion estimation process just when it is needed most. Figure 7.7 shows a comparison between motion estimation on the source and motion estimation on the reconstructed image based on an MPEG-2 simulation. As the bit rate is reduced, the reconstructed image diverges more and more from the source. As a result, motion vectors derived from the source produce less and less accurate predictions, thus reducing the picture quality and PSNR, as shown in Figure 7.7.

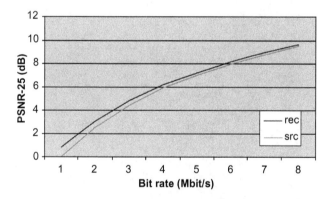

Figure 7.7   Comparison between motion estimation on source and motion estimation on the reconstructed image

A further potential hazard of hierarchical motion estimation arises from the need to generate a hierarchy of downsized images. If short filters are used for this operation, the small images contain an increasing amount of aliasing in each layer of the hierarchy. Aliasing components of detailed images, however, tend to move at different speeds and in different directions, compared with the source image, thus fooling the motion estimation process in the higher layers. Figure 7.8a–d shows the four layers of hierarchy generated with relatively short down-sampling filters. It can be seen that the pictures generated with short filters contain significant alias components.

Figure 7.8   (a) Source image at SDTV resolution. Same image down-sampled to (b) CIF resolution, (c) QCIF resolution and (d) QQCIF resolution

While the aliasing effects can be reduced by using better down-sampling filters, there is a more fundamental problem with hierarchical motion estimation. Consider a two-stage hierarchical motion estimation system analysing an SDTV sequence, e.g. the one shown in Figure 7.8a with a horizontal pan of just one pixel per frame. Down-sampled to CIF resolution, a horizontal shift of one pixel per frame cannot be detected, because neither a vector of (0,0) nor one of (2,0) gives a good match. Chances are that the motion estimation stage at CIF resolution will choose a vector far away from the (1,0) result, if similar patterns are found within the search range. For example, a better match could be found in another window of the skyscraper. The same applies to horizontal (and also vertical) motion displacements of 3, 5, 7, etc. pixels per frame. Once the wrong starting vector is chosen, it is sometimes difficult to recover from a sub-optimum starting point at a later stage.

7.2.3   Other block-matching methods

Although most of the disadvantages and potential problems of hierarchical motion estimation can be overcome by careful design, a large number of other block-matching techniques have also been developed. The aim is to achieve almost the same performance as exhaustive search but with less processing power.

7.2.3.1   Spiral search

One such method is the so-called spiral search [11]. Whereas in an exhaustive search the order of the search pattern is irrelevant since all search positions are examined irrespective of previous results, a spiral search is based on the assumption that once a certain match criterion has been achieved close to zero motion, a further search is no longer necessary. The search starts at the motion vector position zero and spirals out to larger and larger vector sizes. As soon as the match criterion has been achieved, the search is terminated. The match criterion itself can be relaxed as the vector size increases. Figure 7.9 gives an illustration of a spiral search.

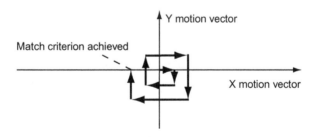

Figure 7.9   Spiral search

Since the probability of finding a good match near zero is extremely high, as can be seen in Figure 7.4, a spiral search is, on average, considerably faster than a full search. The problem is that the processing time for each macroblock and even for an entire picture is not deterministic. Sequences with slow motion are searched quickly whereas the search of fast and/or complex motion takes considerably longer. In real-time encoders a spiral search would need to have a set time limit, which would limit the performance in the most critical sequences where powerful motion estimation is most needed.

7.2.3.2   Sub-sampling method

A simple method of reducing the computational processing requirements of block-matching motion estimation algorithms is to reduce the number of pixels used to calculate the SAD. Instead of calculating the SAD value across all pixels of the macroblock, a sub-set of pixels can be used to calculate the match criterion. For example, by using a quincunx sub-sampling pattern as shown in Figure 7.10 the number of calculations can be reduced by a factor of 2 without reducing the horizontal or vertical motion estimation accuracy on non-interlaced video signals [12]. Unfortunately, on interlaced video signals, a pattern as shown in Figure 7.10 reduces the horizontal accuracy because the quincunx pattern degenerates into a simple horizontal sub-sampling pattern on each of the two fields.

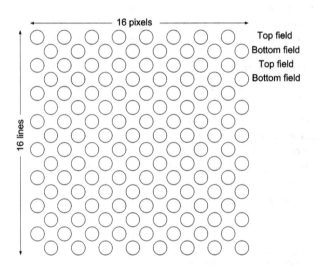

Figure 7.10   Frame-based macroblock sub-sampled in a quincunx pattern

7.2.3.3   Telescopic search

Telescopic search is a method specifically suitable for compression systems using B frames. When using B frames, the search from one P frame to the next has to be considerably larger than without B frames because the search has to be carried out across all intermediate B frames as shown in Figure 4.20. The assumption is that the motion of an object can be tracked through the B frames until the best match is found in the previous P (or I) frame. Since the motion displacement from one frame to the next is relatively small, a smaller search range can be used in each B frame search, thus improving processing efficiency [13]. Figure 7.11 gives a graphical illustration of a telescopic search across two B frames. It can be seen that with the accumulation of small search ranges a large P-to-P frame search range can be achieved.

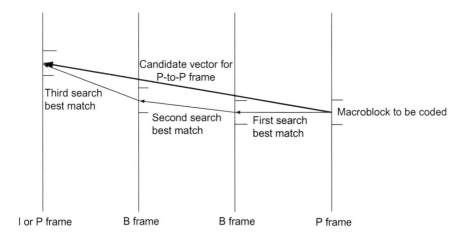

Figure 7.11   Telescopic search across two B frames

7.2.3.4   Three-step search

Other block-matching motion estimation techniques rely on the assumption that the correlation surface within the search range is relatively smooth, that is to say that once a 'good' match has been found, the 'best' match is assumed to be nearby. This is true in many sequences but not in those with periodic patterns such as the windows of the building shown in Figure 7.8.

Rather than down-sampling the video signal into a hierarchy of smaller and smaller pictures, one can start by calculating the match criterion for a relatively small number of test points scattered across the search range in a regular pattern. After finding the best initial match, the search continues in smaller steps in the vicinity of the best initial match. This process can be repeated several times until the desired motion vector accuracy is achieved.

Figure 7.12 gives a graphical example of this search algorithm. The initial search points are indicated as '+'. Once the best initial match has been found, the search continues by testing the refinement points around the best initial match, indicated as '□'. The final motion vector is found by calculating the match criterion for the points near the best refinement match, indicated as '○'. If the search is carried out in three stages, as shown in the example of Figure 7.12, this algorithm is often referred to as a 'three-step search algorithm' [14]. However, there are many different versions of this type of search algorithm [15–17].

7.3   Correlation algorithms in the frequency domain

Block-matching algorithms are the most popular motion estimation methods for video compression. They are relatively simple to implement and are configurable for all shapes and sizes. Nevertheless, there are other more sophisticated approaches to motion estimation. Although motion estimation in the frequency domain is

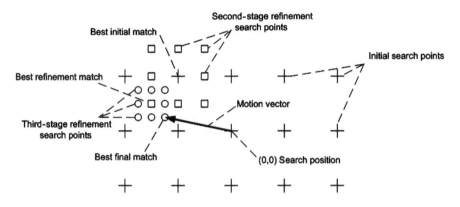

Figure 7.12   Three-stage search algorithm

(so far) rarely used in video compression, it can produce superior results in other applications such as standards conversion – a reason enough to provide a brief overview on such methods.

There are two distinct motion estimation algorithms operating in the two-dimensional spatial frequency domain. Although the two algorithms have many similarities, they are nonetheless quite different in detail. Both algorithms use two-dimensional Fourier transforms to calculate the convolution between search and reference blocks and both use some form of normalisation in the frequency domain, in order to calculate a correlation surface between a reference and a search area (see Appendix F). However, this is where the similarities end. Rather than explaining the two algorithms with complicated mathematical formulae, we will examine the principle using one-dimensional graphical examples.

7.3.1   Phase correlation

The first method, called 'phase correlation', is based on the principle that a displacement in the spatial domain corresponds to a phase shift in the frequency domain. The phase shift is determined by calculating the Fourier transform of co-located rectangular areas of both reference and search pictures. After multiplication of the Fourier transform of the reference block with the complex conjugate Fourier transform of the search block, the resulting spectrum is normalised by dividing each element by its magnitude [18]. This leaves a normalised correlation surface that contains all motion information in the phase values of the spectral components. To obtain the motion vectors, one simply needs to calculate the inverse Fourier transform of the normalised spectrum.

Figure 7.13 gives a graphical example of this method. There are two types of motion in this example. The rectangular object is moving one pixel to the left whereas the triangular object is moving two pixels to the right. After inverse Fourier transform of the normalised spectrum, we can observe two distinct peaks, representing the two types of motion within the search and reference blocks.

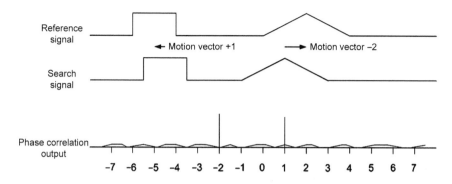

Figure 7.13   Graphical illustration of phase correlation motion estimation

Therefore, rather than calculating a motion vector for each macroblock, phase correlation calculates the number of candidate vectors within a search and reference area, which typically consists of 16 or more macroblocks. The advantage is that several motion vector candidates are calculated in one go. The disadvantage is that the actual motion vector applicable to each macroblock still has to be chosen from the selection of candidate motion vectors. This is typically done with a block-matching algorithm.

7.3.2   Cross-correlation

Cross-correlation methods are frequently used in signal-processing applications, particularly if similarities between two signals are to be detected [19]. Rather than trying to calculate all motion vector candidates within a given search and reference picture area in one go, the cross-correlation method finds the best correlation match between one reference macroblock and the video signal within the search area [20]. For this purpose, the reference picture area, which has to be the same size as the search area in order to perform the cross-correlation in the Fourier domain, is augmented with zeros.

The cross-correlation itself is carried out in a way similar to the phase-correlation method. That is to say that both reference and search blocks are Fourier transformed and the transformed reference block is multiplied with the complex conjugate of the transformed search block. The normalisation algorithm, however, is far more complex in this method than that in the phase-correlation method. However, proper normalisation is necessary in order to avoid false matches with similar objects of higher amplitudes.

Figure 7.14 gives a graphical example of the cross-correlation method. In this method the reference block contains only one macroblock. The rest of the block is set to zero. The cross-correlation operation finds the best match in the search block. If the search block contains an exact match of the reference macroblock, then the correlation surface (after normalisation and inverse Fourier transformation) reaches a maximum of 100 per cent at the match location. In any case, the motion

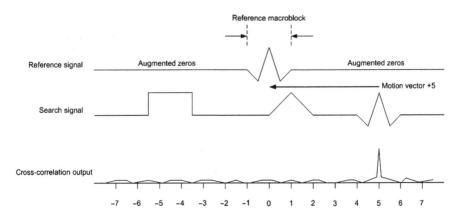

Figure 7.14   Graphical illustration of cross-correlation motion estimation

vector with the best prediction is determined by the global maximum of the cross-correlation output.

The advantage of this method is that it derives the optimum motion vector directly without having to do further block-matching searches [21]. Furthermore, cross-correlation is a better match criterion for compression than basic block matching because it eliminates the DC component.

7.4   Sub-block motion estimation

Up to now we did not concern ourselves with the motion estimation block size. That is to say the motion estimation algorithms considered earlier are applicable to any block size. In MPEG-2 motion compensation blocks are either 16×16 pixels for frame-based or 16×8 pixels for field-based motion compensation. Advanced coding algorithms, however, permit a larger number of different block sizes for inter-frame predictions. As has been pointed out in Chapter 5, MPEG-4 (AVC), for example, allows seven different motion compensation block sizes from 16×16 down to 4×4 pixels. Figure 7.15 illustrates how 16×16 macroblocks can be divided into different sub-blocks in order to make more accurate predictions at the boundaries of moving objects.

To support different motion compensation block sizes, separate motion estimation searches have to be carried out for each sub-block partition. To carry out a full search for all individual sub-blocks is extremely processing intensive. For that reason, search algorithms have been developed to reduce the processing requirements for multiple sub-block motion estimation [22,23]. In many cases it is sufficient to search for sub-block matches in the vicinity of the best macroblock match.

Although smaller block sizes can produce better predictions at object boundaries and in complex, non-translational motion, the use of smaller motion compensation blocks is not always beneficial because smaller blocks require a larger

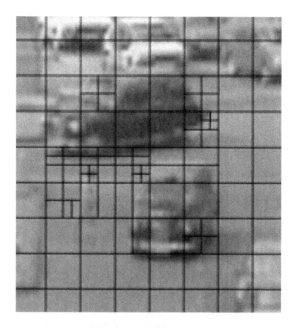

Figure 7.15   Sub-block partitions around moving objects

number of motion vectors. As a result, the coding cost of motion vectors can be higher than that of additional transform coefficients in larger blocks. Once the motion vectors and prediction errors for all sub-block sizes have been found, the best prediction mode can be chosen based on prediction errors as well as the number of bits to code motion vectors and transform coefficients. The larger the number of prediction modes, the more important it is to use RDO methods for mode decisions.

7.5   Sub-pixel motion refinement

The majority of video compression implementations use a full-pixel search algorithm as the first stage of motion estimation, followed by a sub-pixel motion refinement. However, sub-pixel motion estimation can also be carried out in the frequency domain, either using phase correlation [24] or DCT transforms [25]. MPEG-2 requires ½ pixel motion vector accuracy, but in MPEG-4 (AVC) ¼ pixel accuracy is required. Sub-pixel refinement is usually confined to a small area around the best full-pixel match. In some cases motion vectors of neighbouring blocks or macroblocks can be used to accelerate the sub-pixel search [26].

7.6   Concluding remarks

Motion estimation is an important ongoing research topic, not just for video compression but also for many other applications, such as motion measurement [27], time delay estimation [28] and robotic vision [29,30]. It can be treated in a very simplistic way, summing up the differences within a block area, or it can get quite mathematical, using Fourier transforms and correlation techniques. Both types of approaches are actively being researched for many applications. This chapter completes the introduction to the core video compression algorithms. While the next chapter gives a brief introduction on video pre-processing functions, the remainder of this book provides an overview on system aspects of video compression.

7.7   Summary

- Exhaustive motion estimation is a simple yet powerful algorithm, but highly processing intensive.
- Hierarchical motion estimation can cover a vast search range using less processing power, but it does not always achieve the same performance as exhaustive motion estimation.
- Research on motion estimation concentrates on achieving the same performance as exhaustive motion estimation using only a fraction of the processing power.
- Correlation algorithms in the frequency domain are powerful methods for standards conversion and other applications, but they are rarely used for video compression.

Exercise 7.1   Motion estimation search range

Assume that a motion estimator should cover 99 per cent of all translational motion displacements.

- What search range is required in SDTV on P frames using a GOP structure of IBPBP…
- What search range is required in HDTV 1080i with a GOP structure of IBBPBBP…?

References

1. D. Martinez, L. Jae, 'Implicit Motion Compensated Noise Reduction of Motion Video Scenes', IEEE International Conference on Acoustics, Speech, and Signal Processing, April 1985. pp. 375–8.
2. D. Van de Ville, W. Philips, I. Lemahieu, 'Motion Compensated De-interlacing for Both Real Time Video and Still Images', International Conference on Image Processing, Vancouver, September 2000. pp. 680–3.
3. H. Lau, D. Lyon, 'Motion Compensated Processing for Enhanced Slow-Motion and Standards Conversion', International Broadcasting Convention, Amsterdam, September 1992. pp. 62–66.

4. B. Furht, J. Greenberg, R. Westwater, *Motion Estimation Algorithms for Video Compression*, New York: Kluwer Academic Publishers, ISBN 0792397932, 1996.

5. L.-K. Liu, 'Rate-Constrained Motion Estimation Algorithm for Video Coding', *International Conference on Image Processing*, Vol. 2, October 1997. pp. 811–4.

6. W.C. Chung, F. Kossentini, M.J.T. Smith, 'Rate-Distortion-Constrained Statistical Motion Estimation for Video Coding', *International Conference on Image Processing*, Vol. 3, October 1995. p. 3184.

7. M. Gharavi-Alkhansari, 'A Fast Motion Estimation Algorithm Equivalent to Exhaustive Search', IEEE International Conference on Acoustics, Speech, and Signal Processing, Salt Lake City, May 2001. pp. 1201–4.

8. F. Essannouni, R.O.H. Thami, A. Salam, D. Aboutajdine, 'A New Fast Full Search Block Matching Algorithm Using Frequency Domain', Proceedings of the 8th International Symposium on Signal Processing and its Applications, August 2005. pp. 559–62.

9. Y.Q. Shi, H. Sun, *Image and Video Compression for Multimedia Engineering: Fundamentals, Algorithms and Standards*, New York: CRC Press Inc, ISBN 0849334918, 1999.

10. F. Lopes, M. Ghanbari, 'Hierarchical Motion Estimation with Spatial Transforms', International Conference on Image Processing, Vancouver, September 2000. pp. 558–61.

11. N. Kroupis, N. Dasygenis, K. Markou, D. Soudris, A. Thanailakis, 'A Modified Spiral Search Motion Estimation Algorithm and its Embedded System Implementation', IEEE International Symposium on Circuits and Systems, May 2005. pp. 3347–50.

12. K.W. Cheng, S.C. Chan, 'Fast Block Matching Algorithms for Motion Estimation', *International Conference on Acoustics, Speech, and Signal Processing*, Vol. 4, May 1996. pp. 2311–4.

13. W. Zhang, 'Low-Latency Array Architecture for Telescopic-Search-Based Motion Estimation', *Electronics Letters*, Vol. 36, Issue 16, August 2000. pp. 1365–6.

14. L. Tieyan, Z. Xudong, W. Desheng, 'An Improved Three-Step Search Algorithm for Block Motion Estimation', 5th Asia-Pacific Conference on Communications and 4th Optoelectronics and Communications, Beijing, October 1999. pp. 1604.

15. C. Zhu, X. Lin, L. Chau, L.M. Po, 'Enhanced Hexagonal Search for Fast Block Motion Estimation', *IEEE Transactions on Circuits and Systems for Video Technology*, Vol. 14, Issue 10, October 2004. pp. 1210–14.

16. C.H. Cheung, L.M. Po, 'Novel Cross-Diamond-Hexagonal Search Algorithm for Fast Block Motion Estimation', *IEEE Transactions on Multimedia*, Vol. 7, Issue 1, February 2005. pp. 16–22.

17. A.M. Tourapis, 'Enhanced Predictive Zonal Search for Single and Multiple Frame Motion Estimation', *Proceedings of SPIE*, January 2002.

18. S. Kumar, 'Efficient Phase Correlation Motion Estimation Using Approximate Normalization', 30th Asilomar Conference on Signals, Systems and Computers, November 2004. pp. 1727–30.
19. G.M. Dillard, B.F. Summers, 'Mean-Level Detection in the Frequency Domain', *IEE Proceedings Radar, Sonar and Navigation*, Vol. 143, Issue 5, October 1996. pp. 307–12.
20. C.-K. Chen, 'System and Method for Cross Correlation with Application to Video Motion Vector Estimator', US 5,535,288 patent, July 1996.
21. V. Argyriou, T. Vlachos, 'Estimation of Sub-pixel Motion Using Gradient Cross-Correlation', *Electronics Letters*, Vol. 39, Issue 13, June 2003. pp. 980–2.
22. W.I. Choi, B. Jeon, 'Hierarchical Motion Search for H.264 Variable-Block-Size Motion Compensation', *Optical Engineering*, Vol. 45, January 2006.
23. G.B. Rath, A. Makur, 'Subblock Matching-Based Conditional Motion Estimation with Automatic Threshold Selection for Video Compression', *IEEE Transactions on Circuits and Systems for Video Technology*, Vol. 13, Issue 9, September 2003. pp. 914–24.
24. V. Argyriou, T. Vlachos, 'A Study of Sub-pixel Motion Estimation Using Phase Correlation', BMVC 2006, Vol. 1, September 2006.
25. U.-V. Koc, K.J.R. Liu, 'DCT-Based Subpixel Motion Estimation', *International Conference on Acoustics, Speech, and Signal Processing*, Vol. 4, May 1996. pp. 1930–3.
26. Z. Wei, B. Jiang, X. Zhang, Y. Chen, 'A New Full-Pixel and Sub-pixel Motion Vector Search Algorithm for Fast Block-Matching Motion Estimation in H.264', 3rd International Conference on Image and Graphics, December 2004. pp. 345–8.
27. V. Argyriou, T. Vlachos, 'Motion Measurement Using Gradient-Based Correlation', International Conference on Visual Information Engineering, July 2003. pp. 81–84.
28. N.M. Namazi, *New Algorithms for Variable Time Delay and Nonuniform Image Motion Estimation*, Bristol, U.K: Intellect Books, ISBN 0893918474, 1995.
29. M. Terauchi, K. Ito, T. Joko, T. Tsuji, 'Motion Estimation of a Block-Shaped Rigid Object for Robotic Vision', IEEE/RSJ International Workshop on Intelligent Robots and Systems, November 1991. pp. 842–7.
30. F. Caballero, L. Merino, J. Ferruz, A. Ollero, 'Improving Vision-Based Planar Motion Estimation for Unmanned Aerial Vehicles Through Online Mosaicing', IEEE International Conference on Robotics and Automation, May 2006. pp. 2860–5.

Chapter 8
Pre-processing

8.1 Introduction

Pre-processing is an important part of any professional compression encoder. It includes vital functions such as noise reduction, forward analysis, picture re-sizing and frame synchronisation. Very often, the pre-processing functions have direct connections with the main compression engine, which is why, in many cases, integrated pre-processing produces better results than external ones, even if external pieces of equipment might be more sophisticated.

Pre-processing is as much a system function as it is an encoder function. It is closely related to most of the concepts discussed in remaining chapters of this book, and in many cases it would be beneficial to introduce the system aspects first before explaining how pre-processing functions can improve the end-to-end performance. Nevertheless, it was felt that since pre-processing is an integral part of encoders, it should be explained in conjunction with compression algorithms rather than towards the end of the book. There are many cross-references in this chapter, and the reader should not hesitate to read some of the other chapters before returning to this one.

8.2 Picture re-sizing

Most compression encoders have the capability of reducing the picture size prior to compression. The main purpose is to reduce bit rate by coding fewer pixels per period of time. Other applications are to change the aspect ratio or convert from a television format, e.g. SDTV, to a computer format, e.g. VGA. Whereas SDTV formats are derived from analogue line standards with non-square pixels, computer-based and HDTV formats always use square pixels. While horizontal re-sizing is relatively straightforward, vertical re-sizing has the added complication of interlacing.

8.2.1 Horizontal down-sampling

To reduce the number of samples per line, we need to limit the horizontal bandwidth of the video signal, in order to avoid aliasing. For example, to reduce the number of samples from 720 to 360, we can use a finite impulse response (FIR) filter with a cut-off frequency of ¼ of the sampling frequency and then drop every other sample. In general, we want to reduce the number of samples per line by a factor of M/N, where M and N are integers. Appendix G provides a brief introduction of the design of polyphase filters.

The basic concept of sampling-rate conversion is well known [1] and can be briefly summarised as follows: The sampling rate of the input signal is first increased by a factor of M by inserting $(M-1)$ zero-valued samples between each existing sample of the input signal. After filtering, the resulting sequence is sub-sampled by retaining only every Nth output sample. Insertion of the zeros shifts the sampling rate but does not alter the spectrum of the input signal. The purpose of the filter is to remove repeat spectra of the input sequence, which would otherwise cause aliasing after sub-sampling. The conversion filter can generally be described as a linear, periodically time-varying system for the purpose of analysis.

In practice, the insertion of zeros (with the resulting increase in sampling rate) and sub-sampling are avoided by calculating only the required output samples, using a parallel structure of M sub-filter sections. These sub-filters, often referred to as polyphase filters, consist of sub-sets of the coefficients of the conversion filter, interleaved by a factor of M. The polyphase filters themselves are time invariant and are easier to analyse. Since all polyphase filters operate in parallel, the output sequence is re-assembled by taking output samples of each of the polyphase filters in turn. Figure 8.1 shows the sampling waveforms for a ¾ down-conversion.

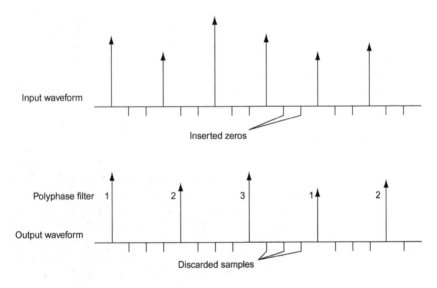

Figure 8.1   Sampling waveforms for a ¾ down-conversion

Although the sampling-rate conversion filter is usually implemented as a parallel structure of polyphase filters, it is initially designed as a conventional low-pass filter by specifying its frequency response, bearing in mind that it operates at M times the input sampling rate. Since the spectrum of the input waveform is periodic in the original sampling rate, the cut-off frequency of the conversion filter needs to be the lower of π/N and π/M in order to avoid aliasing [2]. Any conventional design algorithm may be used [3], but it was found that for relatively short filters, the windowing technique is quite adequate for the initial design. One advantage of the windowing technique is that the pass-band ripple tends to diminish for lower frequencies, which is desirable for video applications, since the visibility of distortion increases with decreasing frequencies.

The prototype conversion filter is then decomposed into polyphase filters and the coefficients are quantised, such that the DC gain of all polyphase filters is unity. The initial DC gain of the polyphase filters is lower, by a factor of M, than that of the prototype filter because of the distribution of the coefficients.

The objective of the filtering process is to remove repeat spectra at multiples of the input sampling rate and to limit the frequency bandwidth to the new Nyquist rate. In doing so, some of the polyphase filters calculate samples at new spatial positions, as shown in Figure 8.1, whereas the position of the output samples of one polyphase filter is set to coincide with input samples in order to keep the picture horizontally aligned. As a result of this phase difference, the frequency responses of the polyphase filters tend to diverge towards the top end of the spectrum if the filters are designed for the full Nyquist rate, as shown in Figure 8.2. This divergence of the polyphase filters can be reduced by lowering the design bandwidth a little [4]. Bearing in mind that the purpose of the down-sampling filters is to reduce compression artefacts, a slight reduction in bandwidth is, in most cases, appropriate.

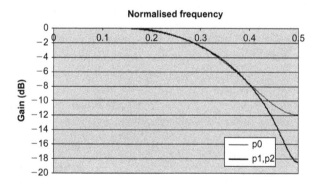

Figure 8.2   Frequency response of polyphase filters for a ³⁄₄ down-conversion

8.2.2   Vertical down-sampling in SDTV systems

Polyphase filters designed in this way are directly applicable to horizontal re-sizing. However, for vertical re-sizing, the interlace structure of television signals needs to be taken into account. In conventional SDTV systems, where both encoder input and decoder output are interlaced, the possibilities for vertical re-sizing are quite limited because most set-top decoders do not support any vertical up-sampling ratios other than 2:1. Moreover, the down-sampling in the encoder has to match the up-sampling method of the decoder to achieve the best results. There are three possibilities for vertical down-conversion, discussed in the following sections.

8.2.2.1   Temporal sub-sampling

With this method the encoder simply drops one field, typically the bottom field, and encodes only the top field. The decoder decodes the top field but presents it twice, in the top and bottom field positions. This is the simplest method and has the advantages that it involves no filtering and preserves a maximum of vertical detail. The disadvantages are 'stair-case' artefacts on diagonal lines and strong temporal aliasing. In fact, temporal aliasing is so strong that moving captions such as ticker tapes are no longer legible when this method is used.

8.2.2.2   Intra-field down-sampling

This method down-samples each field individually using polyphase filters as outlined in the previous paragraphs. It avoids temporal aliasing because no fields are dropped. The disadvantage is the loss of vertical resolution, particularly in the colour difference signals. Bearing in mind that the chrominance signal has already been down-sampled by a factor of 2 due to the 4:2:0 format, further down-sampling by another factor of 2 leaves very little chrominance detail.

8.2.2.3   Vertical-temporal down-sampling

The third method is the most sophisticated and represents a good compromise between vertical and temporal resolutions. It does not drop fields and therefore avoids temporal aliasing and motion judder. Furthermore, by taking adjacent fields into account, this method can achieve a higher vertical resolution than simple intra-field down-sampling. Unfortunately, the corresponding up-sampling method is generally not available in decoder chips because it requires additional field stores and larger FIR filters. Therefore, it has not so far been used in broadcast applications.

Not all applications require matched down- and up-sampling methods. For example, down-sizing from interlaced SDTV to progressive QVGA resolution for mobile or streaming applications is a one-way operation, which does not need to be reversed. Removing the requirement for matched up-sampling opens further possibilities for vertical re-sizing. However, the problem is that these applications generally require progressive scanning formats. This means that we not only need to re-size the image, but also have to de-interlace it.

8.3   De-interlacing

De-interlacing methods are closely related to vertical re-sizing methods of interlaced video signals. For example, dropping the bottom field of an interlaced video signal not only halves the size, but also 'de-interlaces' to a progressive format. If a progressive frame rate of half the field rate of the SDTV picture is sufficient, then this is probably the simplest method of obtaining a smaller, progressive image. In the same way, intra-field or vertical-temporal interpolation filters can be used for de-interlacing.

While the three methods described in Section 8.2 are suitable for de-interlacing, they cannot restore the full vertical resolution of a progressive image. However, this can be achieved by using motion-adaptive [5,6], motion-compensated [7] or iterative [8] interpolation methods. Motion-adaptive methods can attain the full vertical resolution on still or slow-moving objects, whereas motion-compensated methods achieve high vertical detail even on faster-moving objects.

8.3.1   Motion-adaptive de-interlacing

Motion-adaptive de-interlacing techniques use two types of conversion filters: one for static and one for moving areas of the picture. The decision as to which of the two paths to use depends on the amount of movement at or near the current pixel position. A potential problem with motion-adaptive techniques is that switches from one mode to another can be quite noticeable if crude motion detectors are used. To reduce the visibility of switching artefacts, spatio-temporal filters are needed to achieve a gradual transition from one mode to another [9]. Figure 8.3 shows a basic block diagram of a motion-adaptive de-interlacer.

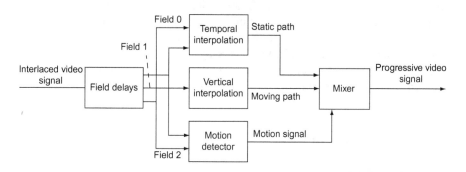

Figure 8.3   Block diagram of a motion-adaptive de-interlacer

The performance of the motion-adaptive de-interlacing process depends largely on the quality of the motion signal. In particular, the visibility of spatio-temporal switching artefacts, accurate mode decisions in the presence of noise or vertical detail and the subjective impression of vertical resolution are all vitally dependent on the proper operation of the motion detector.

8.3.2   Motion-compensated de-interlacing

Motion-compensated de-interlacing techniques are considerably more complex than motion-adaptive techniques. Not only do they need a sub-pixel-accurate motion estimator, but they also require a motion compensator for the interpolation of the missing samples. The benefit of this additional complexity is a higher vertical resolution in moving picture areas [10]. Table 8.1 gives a summary of the advantages and disadvantages of different de-interlacing techniques.

Table 8.1   Summary of de-interlacing algorithms

De-interlacing algorithm	Advantage	Disadvantage
Intra-field filter	Simple, artefact free	Low vertical detail
Vertical-temporal filter	Good compromise between detail and complexity	Slightly more complex due to field delays
Motion-adaptive de-interlacing	High detail in slow moving areas	Loss of vertical detail in fast motion
Motion-compensated de-interlacing	Highest preservation of detail	Most complex algorithm

8.3.3   De-interlacing and compression

In order to assess the picture quality of the entire transmission chain, the aforementioned techniques for de-interlacing have to be evaluated in conjunction with compression. Compression algorithms are quite sensitive to de-interlacing artefacts, insofar as they consume higher bit rates. Furthermore, high picture quality at relatively low bit rates is more readily achieved if the de-interlaced signal is slightly softer in moving areas [11]. Therefore, there is a compromise between detail preservation and reduction of compression artefacts. Both motion-adaptive and vertical-temporal techniques seem very appropriate for de-interlacing prior to compression.

8.4   Noise reduction

8.4.1   Introduction

In analogue video production equipment, noise accumulates as the signal makes its way through the production chain. However, even in digital production studios, camera source noise cannot be avoided. Furthermore, the increased use of compression techniques in the studio means that analogue Gaussian noise is increasingly replaced by digital quantisation noise. Therefore, while the characteristics of video noise are changing over time, the overall noise level is unlikely to drop significantly.

Noise, by its very nature, is unpredictable and usually extends to high frequencies. Therefore, to encode noisy video signals consumes many bits in predictive compression algorithms, be it MPEG-2 or MPEG-4 (AVC). The weighting matrices used in MPEG-2 and in MPEG-4 (AVC) High Profile effectively limit the bandwidth of noise (and of the signal) and thereby reduce to some extent the visibility of the noise. Unfortunately, the presence of the noise makes the encoder work harder, thereby producing more compression artefacts. Effectively, noise is 'converted' to compression artefacts, which might be more objectionable than the original source noise.

If three or more compressed video signals are statistically multiplexed, the effect of video source noise is even more severe than in constant-bit-rate applications. In statistically multiplexed systems (see Chapter 12), noisy video channels claim higher bit rates than equivalent channels with less noise. The bit rate spent on encoding noise is then not available for coding picture information, thus reducing the picture quality of the entire statistical multiplexing group.

8.4.2   Types of noise

Video signals can be impaired by a large variety of different types of noise. Camera noise [12], analogue tape noise, FM 'click' (impulse) noise [13], film grain [14] and quantisation noise [15] from compression equipment are just a few examples. Not only are the levels of noise different between different types of impairment, but the noise spectrum also varies greatly between these types of noise. Each type of noise, therefore, needs to be tackled in its own specific way. Tape noise and FM noise increase towards higher horizontal frequencies, whereas film grain has strong horizontal and vertical mid-band components. Quantisation noise is non-linear and has to be considered as a special case, as discussed in Section 8.4.5.

In the past, expensive pre-processing equipment was necessary to deal with all these types of different noise. Furthermore, the equipment had to be carefully adjusted to the specific type and level of noise present. The availability of large field programmable gate arrays (FPGAs) has made it possible to integrate both the noise analysis and the noise reduction into the front end of today's compression encoders. This provides a cost-effective solution that avoids the need for external noise reduction equipment altogether and makes integration with the rest of the compression equipment easier.

8.4.3   Bit-rate demand of noisy video signals

In statistical multiplexing systems, a number of encoders compete for bandwidth. Without noise reduction, encoders with noisy video signals require more bit rate than those with 'clean' signals. This additional bandwidth to 'encode' noise is required permanently, not just during critical picture scenes, thus reducing the bit rate available to encode picture information on other channels of the statistical multiplexing group. The same argument holds for applications with variable bit rates.

Bit-rate demand is measured by setting the encoder to a constant picture quality, rather than a constant bit rate, and measuring the bit rate required to achieve the desired picture quality. An approximation of fixed picture quality can be achieved by keeping the QP constant.

To find out how much bit-rate saving can be achieved using noise reduction, the bit-rate demand of a number of test sequences has been measured. Since the test sequences have very low noise levels, a moderate amount of Gaussian noise

(40 dB SNR) has been added in order to obtain approximately the same noise level on each test sequence.

Table 8.2 shows the results of this experiment. With constant quantisation, the bit-rate demand of the test sequences with and without noise reduction has been measured. In both cases, the target picture quality, i.e. the QP, of the encoder was held constant. As can be seen, very significant bit-rate savings can be achieved using noise reduction techniques, even with relatively moderate levels of noise.

Table 8.2   Bit-rate demand with 40 dB signal-to-noise ratio

Sequence	Bit-rate demand with 40 dB SNR (Mbit/s)		
	Without NR	**With NR**	**Saving (%)**
Black	1.24	1.07	13.51
Horses (near still)	8.24	4.57	44.51
Tree (still)	5.22	3.52	32.42
Dance (pan)	3.99	3.36	15.90
Susie (talking head)	5.11	4.28	16.31
FTZM (computer graphics)	5.74	5.02	12.66
Table tennis	11.06	8.62	22.09
Popple	13.58	10.76	20.76
Football	11.99	11.03	8.02
Renata	12.47	11.69	6.26
Average	7.86	6.39	19.24

Table 8.2 gives an accurate answer for the bit-rate demand of individual test sequences. However, these results give little indication of the bit-rate saving that can be achieved in practice. A more realistic picture is obtained when bit-rate demand is monitored over long periods of time. Figure 8.4 shows the probability density function (PDF) of the measured bit-rate demand of an encoder set to constant picture quality with and without noise reduction. The analysis covers several hours of material with moderate amounts of noise. As can be seen from Figure 8.4, noise reduction has little effect on the shape of the PDF but moves the entire PDF towards lower bit rates. This means that, for this particular example, a bit-rate saving of 2.2 Mbit/s is achieved on average.

8.4.4   Noise reduction algorithms

There are as many different types of noise reduction algorithms as there are sources of noise. Some have been developed to reduce a particular type of noise [16,17], while others are less specific [18]. The goal of all noise reduction algorithms is to preserve the image detail as much as possible while eliminating the noise. This is usually done with some form of adaptive algorithm, be it motion adaptive, amplitude adaptive, noise-level adaptive or a combination thereof. Algorithmically,

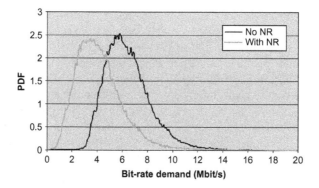

Figure 8.4   Probability density function of bit-rate demand with and without noise reduction

noise reduction methods can be classified into two categories: algorithms working in the temporal domain and those working in the spatial domain.

8.4.4.1   Temporal noise reduction algorithms

Most temporal noise reduction methods are based around a non-linear recursive filter, although in principle non-linear FIR filters could also be used. The filter should have non-linear behaviour because low amplitudes should be treated as noise and therefore should be filtered, whereas high amplitudes should be left untouched [19]. Furthermore, temporal noise reduction algorithms should adapt to motion because noise is most visible in static and slow-moving areas and excessive filtering of moving objects leads to motion blur and/or ghost images.

Hence a simple, yet quite effective, technique is a motion-adaptive temporal recursive filter [20], a basic block diagram of which is shown in Figure 8.5. It uses a motion and amplitude difference detector to control the feedback coefficient. If motion and amplitude differences are low, the control signal selects a high percentage of the frame-delayed signal, thus reducing the noise in this picture area. If motion or amplitude difference is high, the control signal selects the input signal without noise reduction. The transition from one signal to the other is controlled gradually on a pixel-by-pixel basis, in order to avoid switching artefacts.

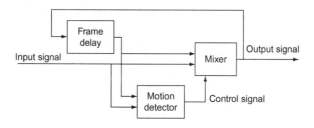

Figure 8.5   Block diagram of a motion-adaptive temporal recursive noise filter

A second, more sophisticated, temporal noise reduction method uses motion compensation techniques in order to enable the feedback loop in moving picture areas too. This can be done using implicit motion compensation techniques for relatively slow-moving areas [21,22] or with explicit motion estimation for faster-moving objects [23]. In practice, very fast-moving picture areas need not be filtered because low-level noise is practically not visible on fast-moving objects. Figure 8.6 shows a block diagram of a noise reduction method with explicit motion estimation. In addition to the frame delay and the motion estimator, this technique requires a motion compensator and a more sophisticated circuit to control the amount of noise reduction.

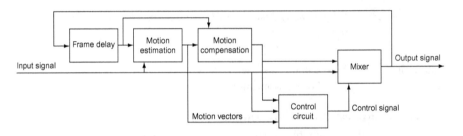

Figure 8.6   Block diagram of a motion-compensated temporal recursive noise filter

8.4.4.2   Spatial noise reduction algorithms

Temporal noise filters, motion-adaptive techniques and motion-compensated techniques all work well for relatively low levels of noise. However, as the noise level increases, temporal filters by themselves are no longer powerful enough to eliminate the noise without introducing motion artefacts. In particular, motion-compensated temporal recursive noise filters can produce irritating effects if the noise level is too high and the temporal noise reduction is set to maximum strength.

For this reason, spatial noise reduction is used in addition to or instead of temporal noise reduction when higher levels of noise are present [24]. In general, the spatial filters are noise-level-controlled non-linear IIR or FIR filters with edge-preserving properties. To achieve best results, the filters operate both in horizontal and in vertical direction. Although these spatial filters preserve high amplitude detail, there is inevitably a trade-off between noise reduction and picture detail/ texture preservation. Consequently, spatial noise reduction filters should be controlled by a noise-level detector.

8.4.4.3   Noise-level measurement

Perhaps even more important than the basic capability to reduce the effects of source noise is the requirement to measure the amount of noise present on the source material [25]. Noise levels can vary over a wide dynamic range from the extremely low levels typically expected at the output of a digital studio to relatively high levels from archive material or some news events. A type of noise that is growing in importance is that caused by upstream compression as used for tape recording, studio processing and digital news gathering. Most noise reduction algorithms, when applied to clean sources at full strength, will give some reduction in picture quality. Operators, however, cannot be expected to change the configuration of encoders for each new type of input source.

8.4.4.4 Impulse noise reduction

In addition to the noise filters described in Section 8.4.4, there are, of course, a plethora of other noise reduction methods. One of the more important ones is a median filter in general [26] or an extremum filter in particular [27]. Extremum filters replace each pixel value with either the minimum or the maximum value of neighbouring pixels, whichever is closer to the mean. Such filters can be used to remove impulse noise.

There are a number of sources that produce impulse noise on video signals, including bit errors on digital lines or tape recorders and FM 'click' noise to name just a few. The high amplitudes of impulse noise prevent the algorithms described in the previous sections from having any effect on impulse noise. Extremum and median filters, on the other hand, are designed to remove these impulses. A more general form of a median filter is a morphology filter [28]. If the threshold below which the morphology filter operates is set too high, then the picture loses texture and acquires a 'plastic' look. However, at low levels, morphology filters can remove residual noise and thus help to reduce bit-rate demands.

8.4.5 De-blocking and mosquito filters

If the bit rate of an MPEG-2 encoder is set too low, critical picture material is likely to suffer from blocking artefacts. Even if the bit rate is quite reasonable, MPEG-2 often suffers from mosquito noise. Mosquito noise is caused by DCT quantisation noise around the edges of objects in front of plain backgrounds. If the quantisation in MPEG-2 is too high, DCT base functions spread into plain areas, causing a noise that looks similar to mosquitos swarming around the object. Figure 8.7 shows an example of mosquito noise and blocking artefacts.

Since MPEG-4 (AVC) encoders are often sourced with video signals that have previously been MPEG-2 compressed (e.g. in turn-around systems), blocking artefacts and mosquito noise are not uncommon at their input. These distortions not only propagate through the MPEG-4 (AVC) system, but also increase the bit-rate demand of the MPEG-4 (AVC) encoder. Therefore, it makes sense to remove the compression artefacts generated by MPEG-2 encoding before re-encoding in MPEG-4 (AVC).

8.4.5.1 De-blocking filters

De-blocking filters basically consist of spatial FIR filters with wide impulse responses for maximum strength. In extreme cases, the FIR filter could be as

Figure 8.7   Example of mosquito noise

wide as the block itself. Several algorithms have been developed for the removal of MPEG blocking artefacts [29,30]. Hence, the problem with de-blocking filters is not so much the removal of block edges; it is the distinction between image edges and block edges caused by compression [31]. In moving areas, block edges are not always aligned with the macroblock structure because motion compensation moves them into different spatial positions. Furthermore, if the previous encoder was not configured to use full resolution, vertical block edges are not macroblock aligned after up-conversion to full resolution. This means that the spatial location of block edges is unknown, even in intra-coded pictures. An added complication is that the quantisation applied to an image area is not known either.

For these reasons, de-blocking filters are much easier to implement as post-processing filters in decoders where information on image resolution and quantisation is readily available, rather than as pre-processing filters in encoders (see Chapter 11). Unfortunately, not many MPEG-2 decoders support de-blocking filters. Even in MPEG-4 (Visual) decoders, where a post-processing filter has been recommended as a non-mandatory option, de-blocking filters are rarely implemented.

8.4.5.2   Mosquito noise filters
The difficulty of implementing compression noise filters in encoders applies equally to mosquito noise filters and to de-blocking filters. Since mosquito noise is more like impulse noise, it can be removed only by strong non-linear filters [32]. Therefore, knowledge of the quantisation is even more important than with de-blocking filters. Otherwise, there is a high likelihood that genuine detail will be filtered out as well.

8.5 Forward analysis

8.5.1 Introduction

High-end real-time encoders analyse the properties of the video input signal at the very front end of the encoder. The more the information that can be extracted from the video signal, the higher the coding efficiency if this information is made use of in other pre-processing functions and/or in the compression engine itself. We have already mentioned in Section 8.4.4.3 that powerful noise reduction circuits are dependent on accurate noise-level measurements in order to make sure that the appropriate level of noise reduction is applied, no matter what video source is used. Similarly, de-blocking and mosquito noise filters can function optimally only if an estimate of the level of MPEG distortion is available. However, in addition to these, there are numerous other parameters to be analysed.

8.5.2 Film mode detection

A considerable proportion of broadcast material is film originated. Since the compression efficiency for film material is much higher than for interlaced material, particularly in 60 Hz countries, it is important to detect the presence of film-originated input and change the compression algorithm accordingly.

For film transfer to 60 Hz video, the well-known 3:2 pull-down method is used to generate 60 Hz video fields from 24 Hz film frames. In this method, three and then two fields are 'pulled down' from successive film frames, thus generating a video sequence that contains two repeated fields in every five video frames. By detecting and eliminating the repeated fields, it is possible to reconstruct and compress the original film sequence, resulting in a significant improvement in coding efficiency [33].

Unfortunately, in reality the situation is somewhat more difficult. Once the film has been transferred to video, the material is usually subjected to further editing in the video domain. As a result, the 3:2 pull-down sequence that would otherwise be readily detectable is disrupted, and reliable detection becomes more involved. Efficient encoders, however, should detect and eliminate repeated fields even in heavily edited material.

The transfer of 24 film frames per second to 25 video frames per second is done either by speeding up the film by 4 per cent or by occasionally inserting additional fields. In either case, most video frames will end up with what is effectively progressive rather than interlaced scanning. By detecting the progressive format and forwarding this information to the compression engine, optimum coding tools can be selected for each format.

8.5.3 Scene cuts, fades and camera flashes

Scene cuts, fades and camera flashes disrupt the temporal correlation of video signals. If the compression engine is unaware of them, it can lead to noticeable distortions. Scene cuts are both relatively easy to detect and most efficiently dealt with by inserting an I frame at the beginning of the new scene.

Camera flashes and other sudden changes in brightness can fool the motion estimation engine and lead to bad predictions in surrounding frames. If possible, it would be advantageous to avoid taking predictions from frames containing flashes or sudden brightness changes.

Fades, and in particular cross-fades from one scene to another, are both difficult to detect and difficult to deal with [34]. Effectively, they have to be treated as highly critical scenes, even if the two scenes themselves are less critical.

8.5.4   Picture criticality measurement

The bit-rate demand of compression encoders depends on spatial as well as temporal redundancies in the video signal. Sequences with high spatial detail and/or fast, complex motion are difficult to encode and, therefore, are often referred to as critical sequences. Advanced warning of changes in criticality can improve the behaviour of statistical multiplexing as well as rate control algorithms [35]. In particular, at scene cuts it is useful to know in advance if the new scene is more difficult to encode than the current scene, less critical or about the same.

8.5.5   Global coding mode decisions

A thorough analysis of the input video signal can not only estimate the criticality of the video signal, but also provide an indication of which coding modes are likely to be most appropriate for the current scene. For example, the pre-processor could determine the GOP length as well as the GOP structure; i.e. how many B frames should be used in between I or P frames [36,37]. Furthermore, by analysing the interlaced video signal in the vertical-temporal domain, it could give an indication whether field or frame picture coding is likely to be more efficient.

8.5.6   Detection of previously coded picture types

In addition to the analysis outlined in the Sections 8.5.2 to 8.5.5, some high-end broadcast encoders are capable of detecting whether or not the video signal has undergone previous MPEG compression [38]. In fact, not only can they detect residual coding artefacts from previous compression engines, but they can also decide whether a particular frame was I, P or B coded. This functionality is useful in turn-around systems where end-to-end performance of concatenated MPEG systems is crucial. Chapter 14 explains this in more detail.

8.6   Concluding remarks

In this chapter we had a look at a number of pre-processing operations that are commonly used in high-end real-time broadcast encoders. As the compression technology itself matures, advances in pre-processing operations can provide further improvements in picture quality. This is particularly true for encoders used in turn-around systems (see Chapter 14), in statistical multiplex systems (see Chapter 12) and in mobile applications (see Chapter 10).

In addition to the video pre-processing functions mentioned in this chapter, there are also other pre-processing functions required in real-time encoders that are not directly related to video processing. Arguably the most important one is the extraction of ancillary data carried in the serial digital interface (SDI). These include high-bandwidth data, such as digital audio [39] carried in the horizontal blanking interval (Horizontal Ancillary Data HANC), and low-bandwidth data, such as time codes [40], camera-positioning information [41], active format description and bar data [42], etc., carried in the vertical blanking interval (Vertical Ancillary Data VANC). Furthermore, pan and scan data, aspect ratio information, teletext and subtitles (or closed captions), carried in the vertical interval of analogue SDTV signals, are extracted and digitised by the pre-processor.

8.7 Summary

- Pre-processing functions can significantly improve the performance of video compression encoders.
- Noise reduction and de-blocking filters not only improve picture quality, but also reduce bit-rate demand.
- Temporal noise reduction, even if it uses motion compensation techniques, is not powerful enough to reduce high levels of camera noise or film grain. Only spatial noise reduction filters can cope with very high levels of noise.
- Forward analysis helps to optimise encoder performance depending on the video input signal.

Exercise 8.1   De-interlacing

An encoder is configured to convert a 525-line SDTV input to CIF resolution using temporal sub-sampling. When 3:2 pull-down film material is used, the decoded video signal exhibits strong motion judder. What is the possible cause of that and how can it be avoided?

Exercise 8.2   Vertical down-sampling

A pre-processor is down-sampling a 576i interlaced signal to a 288p progressive sequence using temporal sub-sampling. What does the down-converted signal look like when the interlaced signal contains

(a) the highest vertical frequency?
(b) the highest temporal frequency?

Exercise 8.3   HD to SD conversion

A pre-processor is converting a 1080i25 signal to a 576i25 signal with an aspect ratio of 4:3. What conversion processes are required?

Exercise 8.4   Encoder optimisation

A test engineer needs to optimise an encoder for a given test sequence. Which of the following parameters can be optimised for visual quality using PSNR measurements?

- noise reduction parameters
- pre-processing filter bandwidth
- motion estimation search range
- GOP length
- number of B frames
- adaptive QP

References

1. R.E. Crochiere, L.R. Rabiner, 'Interpolation and Decimation of Digital Signals – a Tutorial Review', *Proceedings of the IEEE*, Vol. 69, Issue 3, March 1981. pp. 300–31.
2. P. Pirsch, M. Bierling, 'Changing the Sampling Rate of Video Signals by Rational Factors', Proceedings of the 2nd European Signal Processing Conference, Erlangen, 1983.
3. L.R. Rabiner, B. Gold, *Theory and Application of Digital Signal Processing*, Prentice-Hall, New York: Englewood Cliffs, ISBN 0139141014, January 1975.
4. A.M. Bock, 'Design Criteria for Video Sampling Rate Conversion Filters', *Electronics Letters*, Vol. 26, Issue 26, August 1990. pp. 1259–60.
5. C. Philips, A.M. Bock, 'Motion Adaptive Standards Conversion for Broadcast Applications', IEE Colloquium on Interpolation in Images, London, May 1991.
6. S.-F. Lin, Y.-L. Chang, L.-G. Chen, 'Motion Adaptive De-Interlacing by Horizontal Motion Detection and Enhanced ELA Processing', *Proceedings of the 2003 International Symposium on Circuits and Systems*, May 2003. pp. II/696–9.
7. Y.-L. Chang, P.-H. Wu, S.-F. Lin, L.-G. Chen, 'Four Field Local Motion Compensated De-Interlacing', IEEE International Conference on Acoustics, Speech and Signal Processing, May 2004. pp. V/253–6.
8. J. Kovacevic, R.J. Safranek, E.M. Yeh, 'Deinterlacing by Successive Approximation', *IEEE Transactions on Image Processing*, Vol. 6, Issue 2, February 1997. pp. 339–44.
9. A.M. Bock, 'Motion Adaptive Standards Conversion Between Formats of Similar Field Rates', *Signal Processing: Image Communication*, Vol. 6, Issue 3, June 1994. pp. 275–80.
10. M. Biswas, T.Q. Nguyen, 'Linear System Analysis of Motion Compensated De-Interlacing', IEEE International Conference on Acoustics, Speech and Signal Processing, May 2004. pp. III/609–12.

11. A.M. Bock, J. Bennett, 'Multi-Purpose Live Encoding', 3rd International Conference 'From IT to HD', London, November 2006.

12. A. Bosco, A. Bruna, G. Santoro, P. Vivirito, 'Joint Gaussian Noise Reduction and Defects Correction in Raw Digital Images', *Proceedings of the 6th Nordic Signal Processing Symposium*, 2004. pp. 109–12.

13. A.M. Bock, A.P. Gallois, 'Simulation and Modelling of FM Clicks', *IEEE Transactions on Broadcasting*, Vol. BC-33, Issue 1, March 1987. pp. 8–13.

14. S.I. Sadhar, 'Image Estimation in Film-Grain Noise', *IEEE Signal Processing Letters*, Vol. 12, Issue 3, March 2005. pp. 238–41.

15. A.B. Watson, G.Y. Yang, J.A. Solomon, J. Villasenor, 'Visibility of Wavelet Quantization Noise', *IEEE Transactions on Image Processing*, Vol. 6, Issue 8, August 1997. pp. 1164–75.

16. G. Louverdis, I. Andreadis, N. Papamarkos, 'An Intelligent Hardware Structure for Impulse Noise Suppression', *Proceedings of the 3rd International Symposium on Image and Signal Processing and Analysis*, September 2003. pp. 438–43.

17. Y. Huang, L. Hui, 'An Adaptive Spatial Filter for Additive Gaussian and Impulse Noise reduction in Video Signals', *Proceedings of 4th International Conference on Information, Communications and Signal Processing*, December 2003. pp. 523–6.

18. A. Sidelnikov, O. Stoukatch, R. Hudeev, 'Intelligent Noise Reduction Algorithm and Software', IEEE-Siberian Conference on Control and Communications, October 2003. pp. 87–88.

19. J.H. Chenot, J.O. Drewery, D. Lyon, 'Restoration of Archived Television Programmes for Digital Broadcasting', International Broadcasting Convention, Amsterdam, September 1998. pp. 26–31.

20. S.-Y. Chien, T.W. Chen, 'Motion Adaptive Spatio-Temporal Gaussian Noise Reduction Filter for Double-Shot Images', IEEE International Conference on Multimedia and Expo, July 2007. pp. 1659–62.

21. D. Martinez, J. Lim, 'Implicit Motion Compensated Noise Reduction of Motion Video Scenes', IEEE International Conference on Acoustics, Speech and Signal Processing, April 1985. pp. 375–8.

22. O.A. Ojo, T.G. Kwaaitaal-Spassova, 'Integrated Spatio-Temporal Noise Reduction with Implicit Motion Compensation', International Conference on Consumer Electronics, 2001. pp. 286–7.

23. S.L. Iu, E.C. Wu, 'Noise Reduction Using Multi-Frame Motion Estimation, with Outlier Rejection and Trajectory Correction', IEEE International Conference on Acoustics, Speech and Signal Processing, April 1993. pp. 205–8.

24. K. Jostschulte, A. Amer, M. Schu, H. Schoder, 'A Subband Based Spatio-Temporal Noise Reduction Technique for Interlaced Video Signals', International Conference on Consumer Electronics, June 1998. pp. 438–9.

25. C. Hentschel, H. He, 'Noise Measurement in Video Images', International Conference on Consumer Electronics, 2000. pp. 56–57.

26. P.K. Sinha, Q.H. Hong, 'An Improved Median Filter', *IEEE Transactions on Medical Imaging*, Vol. 9, Issue 3, September 1990. pp. 345–6.

27. N. Young, A.N. Evans, 'Spatio-Temporal Attribute Morphology Filters for Noise Reduction in Image Sequences', International Conference on Image Processing, September 2003. pp. I/333–6.

28. R.A. Peters, 'A New Algorithm for Image Noise Reduction Using Mathematical Morphology', *IEEE Transactions on Image Processing*, Vol. 4, Issue 5, May 1995. pp. 554–68.

29. S. Delcorso, J. Jung, 'Novel Approach for Temporal Filtering of MPEG Distortions', IEEE International Conference on Acoustics, Speech and Signal Processing, May 2002. pp. IV/3720–3.

30. S.D. Kim, J. Yi, H.M. Kim, J.B. Ra, 'A Deblocking Filter with Two Separate Modes in Block-Based Videocoding', *IEEE Transactions on Circuits and Systems for Video Technology*, Vol. 9, Issue 1, February 1999. pp. 156–60.

31. P. List, A. Joch, J. Lainema, G. Bjontegaard, M. Karczewicz, 'Adaptive Deblocking Filter', *IEEE Transactions on Circuits and Systems for Video Technology*, Vol. 13, Issue 7, July 2003. pp. 614–19.

32. H. Abbas, L.J. Karam, 'Suppression of Mosquito Noise by Recursive Epsilon-Filters', IEEE International Conference on Acoustics, Speech and Signal Processing, April 2007. pp. I/773–6.

33. C.C. Ku, R.K. Liang, 'Robust Layered Film-Mode Source 3:2 Pulldown Detection/Correction', *IEEE Transactions on Consumer Electronics*, Vol. 50, Issue 4, November 2004. pp. 1190–3.

34. Z. Cernekova, I. Pitas, C. Nikou, 'Information Theory-Based Shot Cut/Fade Detection and Video Summarization', *IEEE Transactions on Circuits and Systems for Video Technology*, Vol. 16, Issue 1, January 2006. pp. 82–91.

35. M. Jiang, X. Yi, N. Ling, 'On Enhancing H.264 Rate Control by PSNR-Based Frame Complexity Estimation', International Conference on Consumer Electronics, January 2005. pp. 231–2.

36. C.-W. Chiou, C.-M. Tsai, C.-W. Lin, 'Fast Mode Decision Algorithms for Adaptive GOP Structure in the Scalable Extension of H.264/AVC', IEEE International Symposium on Circuits and Systems, May 2007. pp. 3459–62.

37. Y. Yokoyama, 'Adaptive GOP Structure Selection for Real-Time MPEG-2 Video Encoding', *International Conference on Image Processing*, Vol. 2, 2000. pp. 832–5.

38. A.M. Bock, 'Near Loss-Less MPEG Concatenation Without Helper Signals', International Broadcasting Convention, Amsterdam, September 2001. pp. 222–8.

39. SMPTE 272M, 'Formatting AES/EBU Audio and Auxiliary Data into Digital Video Ancillary Data Space', 2004.

40. SMPTE 12M, 'Transmission of Time Code in the Ancillary Data Space', 2008.

41. SMPTE 315M, 'Camera Positioning Information Conveyed by Ancillary Data Packets', 2004.

42. SMPTE 2016, 'Vertical Ancillary Data Mapping of Active Format Description and Bar Data', 2007.

Chapter 9
High definition television (HDTV)

9.1 Introduction

After many years of research and several attempts to introduce HDTV into broadcast systems, HDTV has finally hit the consumer market. It took the combination of several factors to make it happen, the most important of these being the availability of large affordable display devices for consumers. A second important factor was the development of HDTV-capable successors to DVD players. Last but not least, the high compression efficiency of MPEG-4 (AVC) makes it possible to transmit HDTV signals at bit rates that are not much higher than SDTV transmissions were in the early years of MPEG-2. In fact, using DVB-S2 forward error correction (FEC) and modulation, together with MPEG-4 (AVC) compression, makes it possible to transmit HDTV signals within a bandwidth equivalent to that of SDTV MPEG-2 compression with DVB-S FEC and modulation [1].

Unlike SD, which consists of just two versions: 625-line 25 frames/s and 525-line 29.97 frames/s, a large number of HDTV formats are in use. Originally, HDTV was defined as four times the number of pixels of SD, i.e. twice the horizontal and twice the vertical resolution [2]. This had the advantage that four fully synchronised pieces of SDTV equipment could process an HDTV signal. The disadvantage was that legacy SDTV formats would be carried forward into HD standards. Once fully integrated HDTV cameras and studio equipment became available, it was therefore decided that it was better to start afresh.

Rather than using a common sampling frequency standard like the 27 MHz sampling rate for SDTV, a common image format was chosen, consisting of $1\,920 \times 1\,080$ pixels for all interlaced video formats and $1\,280 \times 720$ pixels for all progressive formats. This has the advantage that both versions maintain a pixel aspect ratio of 1:1 and are therefore computer compatible, regardless of the frame rate.

However, the simplicity in spatial resolution led to a multiplicity in temporal resolutions. The $1\,920 \times 1\,080$ interlaced format, usually referred to as 1080i, comes at frame rates of 30, 29.97, 25, 24 or 23.976 frames/s and the $1\,280 \times 720$ progressive format, usually referred to as 720p, can have 60, 59.94, 50, 30, 25 or 24 frames/s. To add to the diverseness, ATSC has also specified some lower resolution formats, which are sometimes also referred to as HDTV since they exceed the Main Level limits of MPEG-2. These are 720×576 at 50 progressive frames/s and 704×480 at 60 or 59.94 progressive frames/s. Lower progressive frame rates are also defined in ATSC but they are not classed as HDTV.

9.2   Compression of HDTV

At first sight it would appear that the compression of HDTV is no different from the compression of any other video format, and indeed MPEG-2 and MPEG-4 (AVC) treat HDTV in the same way as SDTV, albeit at a higher-level compliance point. As we look more closely, however, we find there are a number of issues that are quite specific to HDTV.

The first point to note is that HDTV encoders run at a different operating point to SDTV encoders. While SD MPEG-2 encoders would typically achieve a bit rate of 0.4 bit/pixel, MPEG-2 HDTV encoders can easily achieve 0.3 or less bit/pixel. The difference is due to the physically smaller block sizes of HDTV when viewed on a similar sized display. This trend of bit-rate-to-pixel-rate relationship can also be extrapolated down to smaller screen sizes. At CIF resolution, for example, a bit rate of at least 0.5 bit/pixel would be required to achieve the same kind of picture quality, albeit at a lower spatial resolution than in SD.

A second factor that differentiates HD from SD is the correlation between objective and subjective picture quality. Although less noticeable between SDTV and HDTV, there is certainly a trend that the correlation between objective and subjective picture quality reduces with larger image size. Unlike the bit-per-pixel ratio, this has more to do with the physical size of the display device than with the coding block size.

The reason for the divergence between objective and subjective picture quality on larger displays relates to the size of display device in relation to the field of view. Larger displays tend to be viewed more closely than small devices, for example, hand-held devices. The closer one is to the display device, the more one concentrates on certain areas of the picture, rather than the entire image as a whole. HDTV, however, tends to be viewed on larger displays than SDTV. Measuring the MOS on HDTV display devices requires attention models pointing out which areas of the screen a viewer is most likely to look at Reference 3. On smaller displays the entire screen is the focus of attention. As a result of these observations, the psycho-visual model in high-end HDTV encoders has to be more sophisticated than that in SDTV encoders.

9.3   Spatial scalability

The question of how to introduce HDTV services alongside existing SDTV programmes is as old as HDTV itself. Even in the analogue world of multi-plexed analogue components (MAC), HD was going to be introduced as a spatio-temporal scalable extension to the SD MAC system [4]. After MAC and HD-MAC were dropped in Europe, there was little interest in HDTV for a period of time, except in Japan. However, once HD had been proposed for ATSC terrestrial transmissions in the US, the question of how to transition from SD to HD in Europe and other parts of the world became an increasingly important issue [5]. The proposed solution was the Spatial Scalability Profile of MPEG-2.

In MPEG-2, spatial scalability works by down-sampling the HD signal to SDTV, coding and decoding the SD signal and up-sampling it again to the original HD format. At that point the difference between the up-sampled SD and the original HD signal should be small and relatively easy to compress. The SD signal is transmitted as a Main Profile, Main Level bit stream so that it can be decoded by a normal Main Profile, Main Level decoder. The HD difference signal, on the other hand, is transmitted as a Spatial Scalable Profile, High 1440 Level stream. Figure 9.1 shows a block diagram of the system.

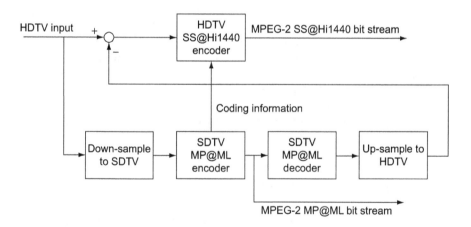

Figure 9.1   Block diagram of an MPEG-2 Spatial Scalable encoder

Unfortunately, there were a number of issues associated with this solution: To start with, it did not support the full 1 920 × 1 088 resolution required for pixels with square aspect ratios. This meant that the solution was not future-proof. Furthermore, simulation tests had shown that the bit-rate saving of scalable video coding compared to simulcast transmission of SDTV and HDTV was relatively small. As a result, broadcasters were not prepared to invest in more expensive transmission and compression equipment. Yet the biggest problem was the complexity of the system. Although theoretical studies on MPEG-2 spatial scalability were carried out [6,7], the small bit-rate savings compared to simulcast and the high encoder and decoder complexity prevented

ASIC decoder and encoder manufacturers from implementing MPEG-2 spatial scalability.

Nevertheless, the development of spatial scalability continued both in MPEG-4 (Visual), where it was refined to fine grained scalability (FGS) [8], and in MPEG-4 (AVC), where it was further developed to today's version of scalable video coding (SVC) [9]. SVC provides a number of useful functionalities: Combined with hierarchical QAM modulation, it can be used for graceful degradation at the edge of the radio frequency (RF) coverage areas [10]. In such a system the SDTV main profile bit stream is transmitted on a simple QPSK (quadrature phase shift keying) modulation, whereas HDTV SVC bit stream is modulated on top of the QPSK modulation. If the received signal at the edge of the coverage area is not strong enough to produce a quasi-error-free bit stream, in other words, if the bit error rate (BER) of the SVC bit stream is too high, then the decoder can switch to the main profile bit stream (which has a lower BER) and display the SDTV instead of the HDTV image. Furthermore, SVC can be used for bit-rate, format and power adaptations [11].

Now that HDTV services have been introduced with MPEG-4 (AVC) coding, SVC is once again being considered for a number of applications. Perhaps one of the most attractive applications is not for HDTV but for transmission to mobile devices [12] (see Chapter 10). Nevertheless, in terms of HDTV applications, SVC could also offer an excellent upgrade path from 720p50/60 to 1080p50/60 or from 1080i25/30 to 1080p50/60. Since it has been shown that SVC of MPEG-4 (AVC) is considerably more efficient than spatial scalability in MPEG-2 [13], and bearing in mind the higher integration density now possible in FPGAs and ASICs, it is likely that SVC applications will be introduced in the near future.

9.4   Progressive versus interlaced

Interlaced scanning is as old as television itself. It was introduced as a band-width-saving technique to provide both high vertical resolution on slow-moving material and high temporal resolution, albeit with less vertical detail. Although interlaced scanning does not seem to fit into today's systems anymore where more and more TV programmes are watched on large, flat-screen or PC displays, one has to bear in mind that a camera with interlaced output, say 1080i25, will have a better signal-to-noise ratio (SNR) than a camera with the same capture device but progressive (i.e. 1080p50) output, even if the video signal is captured on a solid-state device. This is because the interlaced scanning structure can be obtained by filtering the progressively captured images as shown in Figure 9.2. Any low-pass filter, however, improves the SNR.

On the other hand, it would be advantageous if we could finally move towards a processing chain that operates entirely within progressive scanning: from capture, production, distribution right through to the end-consumer display. In terms of picture re-sizing, standards conversion and noise reduction,

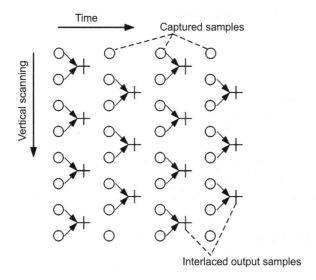

Figure 9.2   Interlaced format derived from a progressive capture device

progressively scanned sequences are easier to process and can ultimately achieve better picture quality. Compression efficiency is also higher on progressive sequences, although it depends very much on how many interlace coding tools are used in the encoder.

There is an argument to say that only 720p should be used for transmission to homes because HDTV will almost always be watched on flat-screen displays and the conversion from an interlaced transmission format to a progressive display is likely to create artefacts. However, most TV display devices are designed for 1080i input and their de-interlacing algorithms have been optimised for the display characteristics. The counter argument is that the up-conversion from the 720p transmission format to the 1080p display format could also produce artefacts. It is mainly when 1080i is decoded directly onto PC displays that interlace artefacts show up. This is because the television concept of interlacing is quite alien to the computer world and most PC rendition software is designed to display interlaced video signals on a frame-by-frame rather than a field-by-field basis with vertical interpolation.

The simple fact of the matter is that at this point in time there is more HDTV production equipment available in 1080i than in 720p, particularly in the 25/50 frames/s format. Furthermore, consumer devices such as Blu-ray support 1080i, and since more and more consumer displays are capable of reproducing the full 1 080 lines per picture height, broadcasters and IPTV companies are reluctant to transmit images with fewer lines per picture height. Therefore, the debate about progressive versus interlaced scanning will probably carry on until the 1080p50/60 format is adopted.

9.5   Compression of 1080p50/60

Although 1080p50/60 has yet to make it to the consumer market, there is an argument that it could and should already be introduced into the production environment. Several European broadcast companies have declared 1080p50 as the ultimate production format [14–16]. Its sampling structure is such that both 1080i25 and 720p50 can be readily derived from it. Furthermore, cameras and production equipment for this format are becoming available, and it is possible to carry a 1080p50/60 signal on a single coaxial cable either in a 3 GHz uncompressed format [17] or in the conventional 1.5 GHz format with mezzanine VC-2 compression [18] (see Chapter 6). The latter has the advantage that the 1080p50/60 signal can be carried on existing 1.5 GHz HDTV networks and routers.

9.6   Bit-rate requirement of 1080p50

To evaluate the bit-rate requirement of 1080p50 as compared to 1080i25, one needs a set of test sequences produced at 1080p50 or a higher resolution format. Fortunately, such test sequences have been produced by Swedish Television (SVT) and made available for research purposes [19]. The production of these test sequences at a resolution of $3\,840 \times 2\,160$ and a frame rate of 50 frames/s made it possible to systematically compare the required bit rates, in terms of MPEG-4 (AVC) compression, of identical material in different HDTV formats [20]. In order to relate the test sequences to general broadcast material, the criticality of the test sequences has to be evaluated.

9.6.1   Test sequences

The criticality of the test sequences can be measured by comparing them against sequences containing a wide range of different content types, including sports, news, movies, etc. To do this, the test sequences have to be down-sampled to a video format in which a large set of reference sequences is available. For the purpose of such a comparison, the bit-rate demand of MPEG-4 (AVC) can be used as a criterion to estimate the coding criticality.

Although the largest set of reference sequences is available in SDTV, a down-sampling ratio of 1.875 would change the spatial criticality significantly. On the other hand, the number of test sequences available in 720p is quite limited. Therefore, a comparison in 1080i25 format represents a reasonable compromise. Of course, there is still no guarantee that the set of reference sequences is representative for broadcast material. However, by making sure that the reference sequences cover a wide range of content types, the criticality of the test sequences can be related to general broadcast material.

To measure the criticality of the test sequences, the bit-rate demand of the test sequences is compared to that of the reference sequences using the same encoder configuration. The encoder should be configured to produce a constant quality, variable bit-rate output corresponding approximately to a grade 4 picture quality as defined in Reference 21. Such a comparison allows the criticality to be expressed in percentage figures. A criticality of 90 per cent, for example, means that 90 per cent of all reference sequences need fewer bits to code the same quality than the test sequence. Table 9.1 gives a summary of the test sequences and their criticality values.

Table 9.1   List of test sequences in the order of coding criticality

Test sequence	Criticality (%)
Park Joy	98.1
Princess Run	97.2
Crowd Run	95.3
Ducks Takeoff	89.7
Seeking	85.0
Passing By	73.8
Tree Tilt	70.1
Umbrella	57.9
Old Town Pan	56.1
Old Town Cross	44.9
Into Castle	30.8
Into Tree	28.0
Dance Kiss	24.3

9.6.2   Test configuration

To carry out the comparison, all test sequences are down-sampled to the conventional 1080i25 and 720p50 HDTV formats, at the full resolution of 1 920 and 1 280 pixels/line, respectively. The less commonly used 1080p25 and 720p25 formats are also included for comparison. In order to investigate the effect of compression efficiency, all test sequences are software encoded in all scanning formats at three different levels of picture quality, corresponding roughly to ITU-R picture grades of excellent, good and fair [22]. The resulting bit rates are related to the corresponding bit rates in 1080i25 format.

Figure 9.3 shows a block diagram of the test set-up. Of the four sample-rate conversion processes shown in Figure 9.3, only the interlace filter and down-sample operation need closer attention. Spatial down-sampling is carried out with conventional polyphase filters designed for the Nyquist cut-off rate, whereas simple frame-dropping is used for the temporal down-sampling.

When interlacing a progressive sequence, there is a trade-off between vertical resolution and interline flicker [23]. Since the interlaced signal undergoes compression, it is not advisable to try and preserve too much detail. Not only would that lead to interline flicker, it would also lead to more compression artefacts. Therefore, a relatively simple intra-field filter was used for this process.

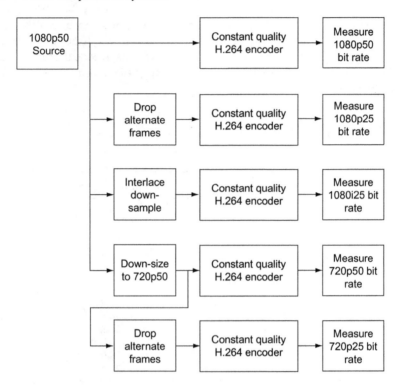

Figure 9.3   Block diagram of the simulation set-up

The reference encoders are configured for constant quality and variable bit rate. Although Figure 9.3 does not show a decoder, the proper operation of the encoders has to be verified to ensure bit-stream compliance. Constant quality encoding is achieved by configuring the encoder for constant quantisation and variable bit rate within the constraints of the buffer model. It should be noted that the constant coding quality is close to, but does not guarantee, constant visual picture quality since it does not take subjective preferences of different scanning formats into account. For example, a 25-progressive frame/s sequence might suffer from motion judder even when its coding quality is high.

9.6.3   Bit-rate results

Figure 9.4 shows the relative required bit rates of 1080p50 as compared to 1080i25. It can be seen that if compressed to a very high picture quality, the 1080p50 format requires significantly more bit rate than 1080i25 although in most cases not as much as the pixel ratio would suggest. Of the 13 test sequences, only one (Old Town Pan) required a higher than the theoretical bit rate at excellent picture quality. As the picture quality target is relaxed, the bit-rate demand drops to a level more closely related to 1080i25 compression.

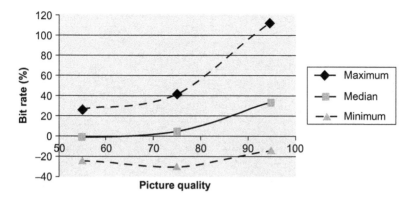

Figure 9.4   Relative bit rate of 1080p50 compared to 1080i25

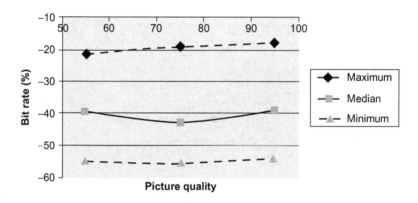

Figure 9.5   Relative bit rate of 720p50 compared to 1080i25

The bit-rate demand of the 720p50 format, as shown in Figure 9.5, is far less dependent on the picture quality target. It remains significantly below the theoretical value of 11 per cent bit-rate saving relative to 1080i25.

1080p25 has the same pixel rate as 1080i25, but requires on average 30 per cent less bit rate, as shown in Figure 9.6. The bit-rate saving of 720p25, shown in Figure 9.7, is very close to the theoretical value of 56 per cent. Neither format shows significant dependency on picture quality. One should bear in mind that while 25 frames/s might be suitable for some programme material, e.g. movies, it can lead to motion judder on medium- to fast-moving content.

9.6.4   Qualification of these results

The results shown above have been achieved using a particular MPEG-4 (AVC) encoder and a set of configurations that are suitable for the various

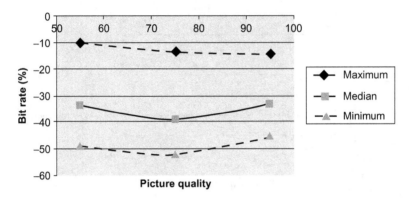

Figure 9.6   Relative bit rate of 1080p25 compared to 1080i25

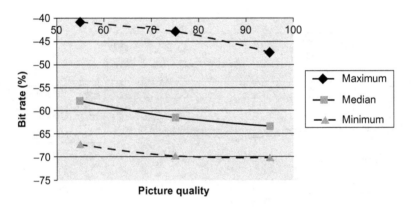

Figure 9.7   Relative bit rate of 720p25 compared to 1080i25

HDTV formats. However, the configurations have not necessarily been fully optimised for each format. Therefore, the results have to be interpreted with some caution. It is entirely possible that an encoder that has been specifically optimised for the 1080i25 format will require a relatively higher bit rate for the 1080p50 format. Conversely, an encoder fully optimised for 720p50, for example, might produce lower bit-rate demands for progressive formats. One has to bear in mind that there are numerous coding tools for interlaced as well as progressive video signals in the MPEG-4 (AVC) standard (see Chapter 5) and that the research to fully exploit these coding tools is still in progress.

9.7   Concluding remarks

Although many coding tools in MPEG-4 (AVC) have been designed with small picture sizes in mind (e.g. ¼ pixel motion vectors, 4 × 4 block sizes, motion vectors outside the picture, etc.), considering the outstanding coding performance of MPEG-4 (AVC) on HDTV material, one could be forgiven for thinking that it had been specifically designed for HDTV. Reducing the bit-rate demand of compressed HDTV signals down to commercially viable values is certainly one of the key factors for the wider deployment of HDTV broadcast and streaming applications, and it would appear that MPEG-4 (AVC) in conjunction with statistical multiplexing is a major factor in achieving just that.

Referring back to the 10 bit 4:2:2 Profile of MPEG-4 (AVC), there is now an interesting choice to be made for high-end production, contribution and distribution compression applications, namely between 10 bit accuracy to avoid posterisation, 4:2:2 Profile or the progressive 1080p50/60 formats. The answer is probably dependent on resolution. Whereas SD will require the 4:2:2 format to allow accurate post-processing (e.g. chroma-key processing) after decompression, HDTV is more likely to move towards the 1080p50/60 format before the need for 4:2:2 processing becomes apparent, bearing in mind that 1080p provides a higher vertical-temporal chroma resolution than 1080i. Due to the fact that gamma correction [24] on large TFT LCD displays tends to emphasize posterisation, it is likely that 10 bit coding will be introduced in contribution and distribution systems and eventually also in DTH systems to avoid such artefacts.

Although the current HDTV format is already producing an excellent picture quality for home cinema, the next generation of an even more impressive video format is currently being developed [25]. It is called the Super Hi-Vision (SHV) format and it consists of $7\,680 \times 4\,320$ pixels at 60 progressive frames/s. This corresponds to 32 times the pixel rate of 1080i and requires a bit rate of 140 Mbit/s in MPEG-4 (AVC).

9.8 Summary

- HDTV encoders achieve higher compression ratios than SDTV because compression artefacts are smaller on HDTV displays, i.e. they are less noticeable.
- SVC provides an excellent upgrade path from 1080i25 to 1080p50 or 1080i30 to 1080p60.
- The production format for HDTV is likely to move from 1080i25/30 to 1080p50/60, but the DTH transmission format will remain at 1080i25/30 and 720p50/60 for the time being.

Exercise 9.1 HDTV contribution

A broadcast company is planning to upgrade their mobile DSNG HDTV equipment from MPEG-2 to MPEG-4 (AVC). To future-proof the investment, they would like to change the production and contribution format from 1080i25 to 1080p50. Do they need to allow for a higher bit-rate contribution link or is the bit rate used for 1080i25 MPEG-2 sufficient?

References

1. S. Bigg, D. Edwards, 'Application of Emerging Technologies to Cost-Effective Delivery of High Definition Television over Satellite', International Broadcasting Convention, Amsterdam, September 2005.
2. M. Haghiri, F.W.P. Vreeswijk, 'HD-MAC Coding for Compatible Bandwidth Reduction of High Definition Television Signals', IEEE International Conference on Consumer Electronics, June 1989. pp. 140–1.
3. K. Ferguson, 'An Adaptable Human Vision Model for Subjective Video Quality Rating Prediction among CIF, SD HD and E-Cinema', International Broadcasting Convention, Amsterdam, September 2007.
4. M. Haghiri, 'HD-MAC: European Proposal for MAC Compatible Broadcasting of HDTV Signal', IEEE International Symposium on Circuits and Systems, May 1990. pp. 1891–4.
5. G.M. Drury, A.M. Bock, 'The Introduction of HDTV into Digital Television Networks', *SMPTE Journal*, Vol. 107, August 1998. pp. 552–6.
6. Y.-F. Hsu, Y.-C. Chen, C.-J. Huang, M.-J. Sun, 'MPEG-2 Spatial Scalable Coding and Transport Stream Error Concealment for Satellite TV Broadcasting', *IEEE Transactions on Broadcasting*, Vol. 44, Issue 2, June 1998. pp. 77–86.
7. T. Shanableh, 'Hybrid M-JPEG/MPEG-2 Video Streams using MPEG-2 Compliant Spatial Scalability', *Electronics Letters*, Vol. 39, Issue 23, November 2003. pp. 1644–6.
8. B.B. Zhu, C. Yuan, Y. Wang, S. Li, 'Scalable Protection for MPEG-4 Fine Granularity Scalability', *IEEE Transactions on Multimedia*, Vol. 7, Issue 2, April 2005. pp. 222–33.
9. H. Schwarz, D. Marpe, T. Wiegand, 'Overview of the Scalable Extension of the H.264/AVC Video Coding Standard', *IEEE Transactions on Circuits and Systems for Video Technology*, Vol. 17, Issue 9, September 2007. pp. 1103–20.
10. N. Souto, 'Supporting M-QAM Hierarchical Constellations in HSDPA for MBMS Transmissions', Mobile and Wireless Communications Summit, July 2007.
11. T. Wiegand, G.J. Sullivan, J.-R. Ohm, A.K. Luthra, 'Introduction to the Special Issue on Scalable Video Coding – Standardization and Beyond', *IEEE Transactions on Circuits and Systems for Video Technology*, Vol. 17, Issue 9, September 2007. pp. 1099–102.
12. T. Schierl, C. Hellge, S. Mirta, K. Gruneberg, T. Wiegand, 'Using H.264/AVC-Based Scalable Video Coding (SVC) for Real Time Streaming in Wireless IP Networks', IEEE International Symposium on Circuits and Systems, May 2007. pp. 3455–8.
13. K. Ugur, P. Nasiopoulos, R. Ward, 'An efficient H.264 Based Fine-Granular-Scalable Video Coding System', IEEE Workshop on Signal Processing Systems Design and Implementation, November 2005. pp. 399–402.

14. A. Quested, 'The HDTV Production Experience', IEE Conference from IT to HD, London, November 2005.
15. P. Bohler, 'HDTV Production Formats', IEE Conference from IT to HD, London, November 2005.
16. H. Hoffmann, T. Itagaki, D. Wood, 'Quest for Finding the Right HD Format', International Broadcasting Convention, Amsterdam, September 2007. pp. 291–305.
17. M. Sauerwald, '3G: The Evolution of SDI', IEE International Conference 'From IT to HD', November 2005. pp. 3/1–17.
18. T. Borer, 'Open Technology Video Compression for Production and Post Production', International Broadcasting Convention, Amsterdam, September 2007.
19. H. Hoffmann, T. Itagaki, D. Wood, A.M. Bock, 'Studies on the Bit Rate Requirements for a HDTV Format with 1920×1080 Pixel Resolution, Progressive Scanning at 50 Hz Frame Rate Targeting Large Flat Panel Displays', *IEEE Transactions on Broadcasting*, Vol. 52, Issue 4, December 2006. pp. 420–34.
20. A.M. Bock, 'Bit Rate Requirements of Different HDTV Scanning Formats', *Electronics Letters*, Vol. 42, Issue 20, September 2006. pp. 1141–2.
21. ITU-R BT.500-11, 'Methodology for the Subjective Assessment of the Quality of Television Pictures', International Telecommunication Union, Geneva, Technical Report ITU-R BT.500-11, 2003.
22. J. Amanatides, D.P. Mitchell, 'Antialiasing of Interlaced Video Animation', *Computer Graphics*, Vol. 24, No. 3, August 1990. pp. 77–85.
23. P.-M. Lee, H.-Y. Chen, 'Adjustable Gamma Correction Circuit for TFT LCD', IEEE International Symposium on Circuits and Systems, May 2005. pp. 780–3.
24. S. Sakaida, K. Iguchi, N. Nakajima, Y. Nishida, A. Ichigaya, E. Nakasu, *et al.*, 'The Super Hi-Vision Codec', IEEE International Conference on Image Processing, September 2007.
25. S. Sakaida, K. Iguchi, N. Nishida, A. Ichigaya, E. Nakasu, M. Kurozumi, *et al.*, 'The Super Hi-Vision Codec', IEEE International Conference on Image Processing, Vol. 1, September 2007. pp. I/21–4.

Chapter 10
Compression for mobile devices

10.1 Introduction

The requirements of video transmission systems to and from mobile devices differ considerably from those of satellite or terrestrial direct-to-home (DTH) transmissions. The main considerations are to keep the power consumption of mobile devices as low as possible and to make sure that frequency deviations due to the Doppler effect in moving receivers do not degrade the performance of the transmission system.

In terms of video compression algorithms for mobile devices, there are also different requirements compared to traditional DTH applications. Not only should the compression algorithm provide high efficiency on relatively small, non-interlaced images, but it should also require little processing power and memory space for encoding as well as decoding.

10.2 Compression algorithms for mobile applications

Since MPEG-2 was designed for interlaced video of SD or HD, it was never seriously considered for mobile applications. One of the first internationally standardised algorithms suitable for mobile applications was MPEG-4 Part 2 (Visual) [1]. This standard included some new coding tools, particularly relevant to smaller images and eliminated some of the MPEG-2 overheads. Optimised for smaller images, MPEG-4 Part 2 introduced a number of new coding tools and achieved a significant amount of bit-rate saving on small images (see Chapter 5 and References 2 and 3). More powerful coding tools were introduced in SMPTE 421M (VC-1), MPEG-4 Part 10 (AVC) and other algorithms, specifically developed for mobile applications [4]. These algorithms achieved substantial bit-rate savings compared to MPEG-2. A comparison between some of the compression tools of these algorithms is shown in Table 10.1. For a more detailed comparison of coding tools, also see Table 6.1.

As far as encoding on mobile devices is concerned, the requirements are not dissimilar to those on wireless cameras. Apart from noise reduction, there are not many pre-processing functions required because the input signal is

Table 10.1   *New coding tools available in advanced video compression algorithms*

MPEG-4 Part 2 Advanced Simple Profile	SMPTE 421M (VC-1) Main Profile	MPEG-4 Part 10 Baseline Profile
¼ pel motion vectors	¼ pel motion vectors	¼ pel motion vectors
Intra-predictions	Intra-predictions	Many intra-predictions
Motion compensation outside of image	Motion compensation outside of image	Motion compensation outside of image
8×8 predictions	8×8 predictions	8×8, 8×4, 4×8 and 4×4 predictions
—	In-loop filter	In-loop filter
VLC	Adaptive VLC	CAVLC
8×8 DCT	Adaptive integer transform	4×4 integer transform

always from the same capture device. In fact, video compression on mobile devices is somewhat simpler than compression on wireless TV cameras because the video source is usually smaller and non-interlaced. The main concern about compression algorithms on mobile devices is low power consumption.

When it comes to TV compression for mobile devices, however, the requirements are somewhat different. Not only the video source has to be re-sized and de-interlaced, but the coding tools themselves have to be carefully selected in order to keep the power consumption on the mobile decoder to a minimum. The following paragraphs explain the trade-offs between compression tools and bandwidth efficiency on the one hand and decoding processing power on the other. However, first of all we need to have a look at the pre-processing requirements.

10.3   Pre-processing for TV to mobile

In addition to the core coding tools shown in Table 10.1, video compression for mobile devices requires a number of sophisticated pre-processing operations in order to make television signals suitable for display on mobile devices. In addition to state-of-the-art pre-processing operations such as noise reduction, de-blocking filters and forward analysis, etc., as explained in Chapter 8, pre-processing for TV to mobile applications involves picture re-sizing and/or cropping in order to make the content suitable for smaller display devices. In fact, one could argue that video compression for mobile devices relies more heavily on pre-processing than on DTH applications [5,6]. This is because the video signal is typically generated in an interlaced format, either in SD or in HD, and therefore not only has to be down-converted and/or cropped to a much lower resolution but also has to be de-interlaced. Whereas horizontal re-sizing can be carried out using simple polyphase FIR filters, vertical re-sizing (and de-interlacing) of interlaced video signals requires a more sophisticated approach, if conversion artefacts are to be avoided (see Chapter 8). Furthermore, mobile services require pixel-accurate image re-sizing, e.g. from SDTV to VGA, since there is generally no up-scaling or post-processing in mobile decoders.

10.4   Compression tools for TV to mobile

MPEG-4 (AVC) Baseline Profile has been proposed as a compliance point for the transmission of video services to mobile devices. Table 10.2 shows the differences between coding tools in Baseline Profile and Main Profile in MPEG-4 (AVC). As far as coding efficiency is concerned, the pertinent differences are that CABAC, B frames and field/frame adaptive coding are not supported in Baseline decoders. Since mobile devices use progressive scanning, field/frame adaptive coding is not relevant in this context. In this section we will analyse the coding benefit of CABAC and B frames at different resolutions, as well as their effect on decode efficiency.

The coding benefit of different compression tools varies considerably, depending on source material, and should therefore be expressed in terms of statistical parameters. In order to obtain realistic and repeatable results, measurements have to be carried out over a large number of test sequences. For each test sequence and picture size, the bit-rate savings of CABAC, the use of B frames and field/frame coding were measured against the Baseline Profile. The results were categorised into percentage bit-rate savings in order to illustrate the full statistical results of the experiment.

Table 10.2   Coding tools supported in MPEG-4 (AVC) Baseline and Main Profiles

Coding tool	Baseline Profile	Main Profile
I and P frames	Yes	Yes
B frames	No	Yes
Multiple reference frames	Yes	Yes
In-loop filter	Yes	Yes
CAVLC	Yes	Yes
CABAC	No	Yes
Flexible macroblock ordering	Yes	No
Arbitrary slice ordering	Yes	No
Redundant slices	Yes	No
Picture-adaptive field/frame coding	No	Yes
Macroblock-adaptive field/frame coding	No	Yes

10.4.1   CABAC

Figure 10.1 shows the histogram of the coding benefit of CABAC entropy coding compared to CAVLC, expressed in terms of bit-rate saving. It can be seen that the highest bit-rate savings are in SD format where the highest percentage of sequences achieve a bit-rate saving between 9 per cent and 11 per cent, with a small number of sequences achieving more than 20 per cent. As the image size is reduced, the bit-rate saving drops to a median value of 7 per cent for CIF and QVGA and down to 6 per cent for QCIF. However, even in QCIF, a small number of sequences achieve a bit-rate saving of more than 12 per cent.

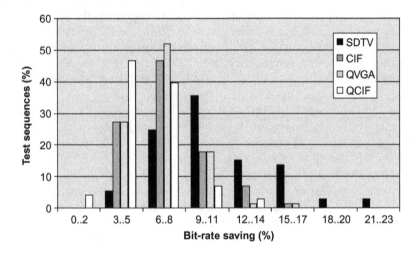

Figure 10.1   Bit-rate saving with CABAC compared to Baseline Profile

10.4.2   B frames

In statistical terms, the bit-rate saving due to B frames is even more impressive. On average, B frames give a bit-rate saving of 17 per cent and up to 40 per cent on some sequences. Furthermore, the bit-rate saving distributions are similar across all resolutions, as can be seen in Figure 10.2.

10.4.3   Main Profile

Since field/frame coding is not to be considered in the context of TV to mobile applications, the coding difference between Main Profile and Baseline Profile consists of the combination of B frames and CABAC entropy coding. Figure 10.3 shows that the average bit-rate savings are 40, 23, 23 and 22 per cent for SDTV, CIF, QVGA and QCIF, respectively. In SDTV, up to 65 per cent bit-rate saving can be achieved in some cases, whereas at lower resolutions, the maximum bit-rate saving is about 50 per cent.

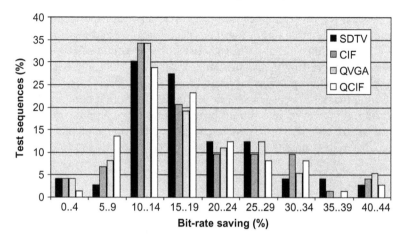

Figure 10.2   Bit-rate saving with B frames compared to Baseline Profile

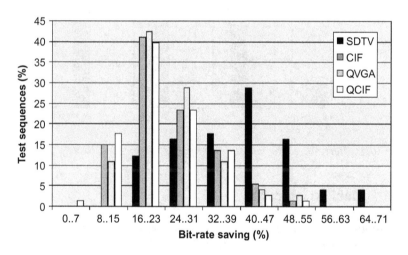

Figure 10.3   Bit-rate saving with Main Profile compared to Baseline Profile

10.4.4   Decode time

Since decode processing power requirements are not only implementation dependent but also content dependent, it is difficult to measure the additional processing power requirements of the coding tools discussed in the previous paragraphs [7]. However, using a software decoder, an estimate of the relative processing requirements can be obtained.

To obtain an estimate of relative decoder processing time, experiments were carried out on a number of test sequences. Figure 10.4 shows the results.

It can be seen that B frame decoding requires between 7 per cent and 11 per cent more decode time than Baseline decoding, depending on resolution. CABAC, on the other hand, requires between 10 per cent and 18 per cent more decode time. Interestingly, the percentage decode time for CABAC reduces with higher resolutions, whereas that for B frames increases. Main Profile decode time increases between 10 per cent and 21 per cent compared to Baseline decoding. However, one should bear in mind that computational processing time is highly platform dependent and different implementations can produce different results.

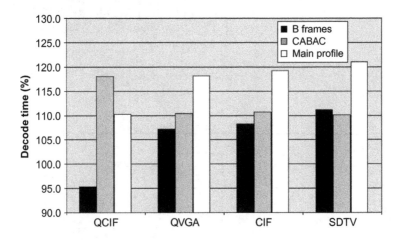

Figure 10.4   Percentage decode time compared to Baseline Profile

10.5   Scalable video coding (SVC)

In addition to the coding tools described in the previous sections, SVC has been proposed for mobile applications [8,9]. SVC provides a range of applications. There are three types of scalability: temporal [10], spatial [11] and SNR [12], sometimes also referred to as quality scalability. With temporal scalability, the base layer codes at a lower frame rate, for example 12.5 frames/s, whereas the upper layer provides the missing frames in between. This could be done by coding an IPP bit stream at Baseline Profile and a BBB stream at the scalable extension, as illustrated in Figure 10.5. By coding the B frames into a separate bit stream, a Baseline decoder can decode the 12.5 frames/s bit stream whereas an SVC-compliant decoder can combine the two bit streams and reconstruct the full 25 frames/s video signal.

With SNR scalable coding, the base layer is coded at a relatively low bit rate using Baseline or Main Profile. Furthermore, the scalable extension encoder encodes the difference between the coarsely quantised base layer and a

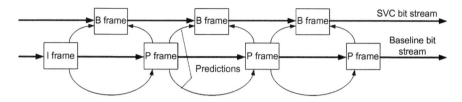

Figure 10.5 Example of temporal scalability from 12.5 frames/s to 25 frames/s

finer quantised version of the video signal. The base layer can be decoded by a conventional Baseline or Main Profile decoder. However, to decode the higher-quality bit stream requires an SVC-compliant decoder.

With spatial scalability, the base layer is coded at a lower resolution (e.g. QVGA) and the SVC encoder encodes the difference between the up-converted base layer and the original higher resolution video signal (e.g. VGA). The principle of spatial scalability has been explained in more detail in the previous chapter (see Figure 9.1). In order to be a viable solution, the combination of the Baseline (QVGA) and SVC bit streams should provide a bit-rate saving compared to a simulcast transmission (QVGA + VGA). This is illustrated in Figure 10.6. In the example shown in the figure, the QVGA signal is encoded at 100 kbit/s. As can be seen, the bit-rate saving of SVC depends on the ratio of the base layer and enhancement layer bit rate. The higher the base layer bit rate compared to the enhancement layer bit rate, the more efficient the SVC encoding compared to simulcast transmission.

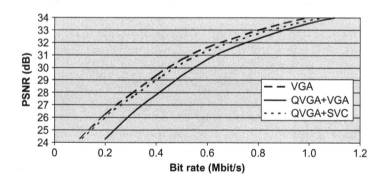

Figure 10.6 Bit-rate saving of SVC compared to simulcast transmission

10.6 Transmission of text to mobile devices

TV services are often accompanied with text information either superimposed on the video signal or as a separate data service. After an SDTV image is re-sized to QVGA size, for example, superimposed text would be no longer legible. Even down-sampling to CIF resolution can cause problems with small text. Furthermore, moving captions become badly distorted if the temporal sampling rate is dropped from 50 fields per second to 25 progressive frames per second. Figure 10.7 shows an example of an SDTV picture with captions down-converted to QCIF resolution. Although the captions are fairly large, they are hardly legible in QCIF resolution. Therefore, small text information has to be transmitted as a separate data service to mobile devices. The receiving device can then overlay the textual information as required.

Figure 10.7   SDTV picture with captions down-converted to QCIF resolution

10.7   Concluding remarks

This chapter has shown that encoding for mobile devices is not simpler, but more complex, than DTH coding at SDTV. While most of the broadcast issues are common, the reduced resolution, low processing power in the decoder and limited bandwidth place additional constraints on the system design. By comparing a number of compression tools, in terms of both coding benefit and decode processing requirements, it has been shown that some coding tools are more appropriate for mobile systems than others. For example, it has been shown that the use of B frames, in particular, can provide substantial bit-rate savings, while their impact on decode processing requirements is relatively moderate. Moreover, video pre-processing and statistical multiplexing techniques reduce average bit-rate demand without placing added requirements on decoders.

10.8   Summary

- Compression encoders for mobile services require sophisticated pre-processing functions for de-interlacing and picture re-sizing.
- Although MPEG-4 (AVC) Baseline Profile has been defined for mobile applications, the use of B frames would provide significant bit-rate savings with little effect on decode processing requirements.
- SVC can be used to transmit the same video content simultaneously at different picture sizes, frame rates and bit rates.

Exercise 10.1   Compression for mobile devices

A mobile transmission company considers using mobile devices that are B frame compliant in order to fit more channels into the same bandwidth. How many channels do they need to transmit until the bit-rate saving of B frames allows them to fit an extra channel into the same bandwidth?

References

1. S.M.M.A Khan, F.Z. Yousaf, 'Performance Analysis of MPEG-4 Encoding for Low Bit-Rate Mobile Communications', International Conference on Emerging Technologies, November 2006. pp. 532–7.
2. R.S.V. Prasad, R. Korada, 'Efficient Implementation of MPEG-4 Video Encoder on RISC Core', *IEEE Transactions on Consumer Electronics*, Vol. 49, Issue 1, February 2003. pp. 278–9.
3. J. Bennett, A.M. Bock, 'In-Depth Review of Advanced Coding Technologies for Low Bit Rate Broadcast Applications', *SMPTE Motion Imaging Journal*, Vol. 113, December 2004. pp. 413–8.
4. K. Yu, J. Lv, J. Li, S. Li, 'Practical Real-Time Video Codec for Mobile Devices', International Conference on Multimedia and Expo, July 2003. pp. 509–12.
5. A.M. Bock, J. Bennett, 'Multi-Purpose Live Encoding', 3rd International Conference 'From IT to HD', November 2006.
6. A.M. Bock, J. Bennett, 'Creating a Good Viewing Experience for Mobile Devices', International Broadcasting Convention, Amsterdam, September 2007. pp. 314–23.
7. Z. Wei, K.L. Tang, K.N. Ngan, 'Implementation of H.264 on Mobile Device', *IEEE Transactions on Consumer Electronics*, Vol. 53, Issue 3, August 2007. pp. 1109–16.
8. R. Fakeh, A.A.A. Ghani, M.Y.M. Saman, A.R. Ramil, 'Low-Bit-Rate Scalable Compression of Mobile Wireless Video', *Proceedings TENCON*, Vol. 2, 2000. pp. 201–6.
9. T. Schierl, C. Hellge, S. Mirta, K. Gruneberg, T. Wiegand, 'Using H.264/ AVC-Based Scalable Video Coding (SVC) for Real Time Streaming in

Wireless IP Networks', IEEE International Symposium on Circuits and Systems, May 2007. pp. 3455–8.

10. C. Bergeron, C. Lamy-Bergot, B. Pesquet-Popescu, 'Adaptive M-Band Hierarchical Filterbank for Compliant Temporal Scalability in H.264 Standard', IEEE International Conference on Acoustics, Speech, and Signal Processing, March 2005. pp. 69–72.

11. C.A. Segall, G.J. Sullivan, 'Spatial Scalability Within the H.264/AVC Scalable Video Coding Extension', *IEEE Transactions on Circuits and Systems for Video Technology*', Vol. 17, Issue 9, September 2007. pp. 1121–35.

12. T. Halbach, T.R. Fischer, 'SNR Scalability by Coefficient Refinement for Hybrid Video Coding', *IEEE Signal Processing Letters*, Vol. 13, Issue 2, February 2006. pp. 88–91.

Chapter 11

MPEG decoders and post-processing

11.1 Introduction

Given that MPEG standards provide exact instructions on how to decode MPEG bit streams, one might assume that decoding MPEG bit streams should be reasonably straightforward. However, in practice, things are not all that simple. MPEG standards provide great flexibility, and while encoders can pick and choose which coding tools to use in which case, decoders have to support all coding tools (of the relevant profile) in all combinations. Although there are usually some test bit streams available, they can never cover the full range of sequence dependencies and all possible combinations of coding tools.

To make things a little easier and to improve interoperability between different encoder and decoder manufacturers, DVB and ATSC issued some guidelines on how to use MPEG standards in DTH broadcast applications [1,2]. Not only did they place certain restrictions on the standard, to make MPEG work in a real-time broadcast environment, e.g. to limit channel change time (the time from when the user presses a button to select a different channel until the new channel is displayed on the screen) without affecting picture quality, but they also defined new standards for electronic programme guides (EPG), subtitles (or closed captions) and many other auxiliary, yet essential, broadcast functionalities. This was certainly of great benefit at the initial introduction of MPEG-2 and is now benefiting the transition from MPEG-2 to MPEG-4 (AVC). Nevertheless, decoder compliance tests are necessary in order to ensure interoperability [3].

Despite all these efforts to achieve interoperability, things can still go wrong. A point in case is the lack of support of field pictures in some early versions of MPEG-2 Main Profile decoders. Although field-picture coding has been one of the Main Profile interlace coding tools from the very beginning, some first generation MPEG-2 decoders did not support it. As a result, high-end broadcast MPEG-2 encoders could not make use of all interlace coding tools.

11.2 Decoding process

Before a real-time MPEG decoder can start decoding video pictures, it needs to establish a timing reference. To do this, the decoder needs to extract programme clock reference (PCR) packets from the transport stream (TS) and synchronise its own 27 MHz system clock to the timing reference transmitted from the encoder. To identify the video and PCR packets of the selected programme, the TS de-multiplexer has to extract the programme association table (PAT), programme map table (PMT) and the conditional access table (CAT), in case the video signal is scrambled, from the TS. This is illustrated in Figure 11.1, which shows a block diagram of a generic MPEG video decoder with TS input. In addition to the PCR and video-processing chains, there are, of course, also audio and data-processing chains and many other functional blocks not shown in Figure 11.1.

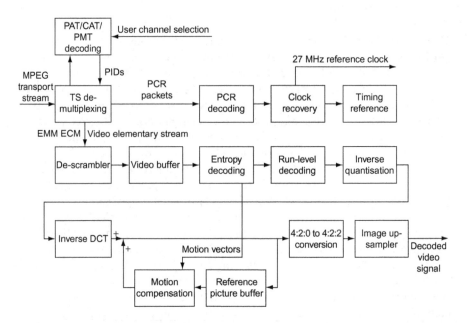

Figure 11.1   Block diagram of a generic MPEG decoder

In case the video signal is scrambled, the decoder needs to obtain entitlement management messages (EMM) and entitlement control messages (ECM) in order to de-scrambler it, otherwise the de-scrambler can be by-passed.

Once the timing reference has been established and the elementary stream PID is known, the video decoder can start looking for sequence and picture headers to initialise the decoding process. When it finds the beginning of an I frame, the decoder starts loading video data into the video buffer. It then waits until the PCR has reached the DTS of the first I frame before it removes the I frame from the video buffer. If there are B frames present, the I frame will be held in the reference picture buffer until the PCR has reached the PTS of the I frame, otherwise it can be presented immediately. Once the decode process has started, compressed video pictures will be removed from the video buffer at regular intervals, e.g. every 40 ms in 625-line 25 Hz video.

Note that the video buffer shown in Figure 11.1 represents three buffers according to the MPEG decoder buffer model [4]: a transport buffer, a multiplex buffer and an elementary stream buffer. MPEG-2 Part 1 defines the sizes of these buffers and the transfer rate between them in order to define buffer compliance. In practice, a decoder does not have to implement three separate buffers in order to be compliant.

After entropy and run-level decoding, the transmitted coefficients are inverse quantised, transformed back into the spatial domain and added to the motion-compensated prediction from previously decoded frames. If the reconstructed images are referenced (I or P) pictures, they are kept in the reference picture buffer for future predictions and display. Referenced pictures are displayed after the relevant B pictures have been decoded, thus reordering the frame structure from transmission order back to display order (see Figure 11.2). Non-referenced (B) pictures do not have to be stored but can immediately be converted back to the 4:2:2 format to be displayed. If the video signal was transmitted at a reduced resolution, it is up-sampled back to full resolution in order to be displayed as a full-resolution SDTV or HDTV signal.

Capture order:
I2 B3 B4 P5 B6 B7 P8 B9 B10 P11 B12 B13 P14 B15 B16

Transmission order:
I2 B0 B1 P5 B3 B4 P8 B6 B7 P11 B9 B10 P14 B12 B13

Display order:
  B0 B1 I2 B3 B4 P5 B6 B7 P8 B9 B10 P11 B12 B13

Figure 11.2   Reordering of B pictures

11.3   Channel change time

There are two types of channel changes: within the same TS and from one TS to another. Channel changing within the same TS is generally faster because PMTs and PIDs are usually already stored in memory and no retuning is required. In this case there are three major factors affecting channel change time: GOP length, i.e. the period of time between intra-coded frames, video buffer size and bit rate.

In DTH broadcast systems the GOP length is usually set to between 0.5 and 1 s, although in MPEG-4 (AVC) systems GOP lengths of more than 1 s are not uncommon. Since the video buffer fills at the speed of bit rate, the maximum delay through the video buffer and therefore the maximum delay from the beginning of the I frame to its presentation is equal to video buffer size divided by bit rate. Although both video buffer size and bit rate can vary over a wide range, the delay through the video buffer is typically less than 0.5 s for MPEG-2 and between 1 and 2 s for MPEG-4 (AVC). Adding up the worst-case numbers for channel change time within the same TS gives a maximum of about 1.5 s for MPEG-2 and 3 s for MPEG-4 (AVC). However, the statistical average is usually just under 1 s for MPEG-2 and just over 1 s for MPEG-4 (AVC).

Channel changing between different TSs, e.g. different digital terrestrial multiplexes or different satellite transponders, can take somewhat longer. However, it is difficult to estimate the channel change time in this case because it is very much implementation dependent.

11.4 Bit errors and error concealment

MPEG-2 and MPEG-4 (AVC) Main Profile and High Profile are based on the assumption that the bit stream to be decoded is quasi error free. Using powerful FEC algorithms, this can be achieved to a large extent on most transmission channels. However, in terrestrial systems, in particular, there are some rare occasions where it is practically impossible to completely avoid bit errors.

When a bit error enters the decoder, there are three stages of recovery:

- In most cases the erroneous bit stream can still be decoded for some period of time, although it produces corrupted video. In other words, the bit stream can still be decoded but the decoded video is not correct. At this stage the decoder is not even 'aware' that a bit error has occurred.
- Sooner or later, however, the corrupted bit stream will produce a syntactic or semantic error and the decoder has to stop decoding because the bit stream can no longer be interpreted.
- The decoder then waits until the next byte-aligned start code, e.g. the next slice header, arrives. At that point the decoding can continue, but, of course, the damage is already done.

In terms of the visual effect of bit errors, there are two possibilities. If the error occurs in a referenced frame (I, P or referenced B frame), it will propagate to all future frames of the current GOP. This is because direct or indirect predictions are taken from the corrupted frame until the next I frame clears the error propagation. If the error occurs in a non-referenced B frame, the distortion is temporally limited to just one frame. Since most MPEG encoders use two or more B frames between referenced frames, one might be misled into thinking that most bit errors should not propagate to other frames. Unfortunately, this is not the case because one has to bear in mind that a large percentage of the total bit rate is spent on I and P frames and only a small percentage of bit rate is spent on non-referenced B frames. Therefore, the likelihood that bit errors occur in non-referenced B frames is rather small. The same is true for MPEG-4 (AVC) bit streams.

Given the GOP structure and the relative sizes of I, P and B frames, it is possible to calculate the average error propagation time, as shown in Appendix H. Figure 11.3 shows how the error propagation time decreases as the relative sizes of P and B frames increase. Sequences with low motion and high spatial detail have relatively small P and B frames compared to I frames and long error propagation times. More critical sequences have larger P and B frames and shorter error propagation times. The example shown in Figure 11.3 is based on a double B frame structure, assuming that B frames are approximately half the size of P frames.

Note that Figure 11.3 is based on the assumption that a small number of bit errors occur within a single compressed video frame. This may be true in satellite and cable systems but is unlikely to apply to digital terrestrial systems. In digital terrestrial systems, errors are likely to occur in longer lasting bursts due to local interference [5].

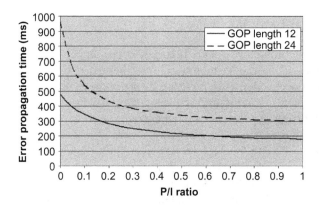

Figure 11.3   Error propagation time as a function of P-to-I frame size ratio

Once a bit error has occurred, the decoder may try to conceal the error. Error concealment is a proprietary operation since MPEG makes no recommendation as to how to deal with bit errors. Many different concealment algorithms have been developed [6–8]. The simplest and most common concealment algorithm is to replace the erroneous or missing macroblocks with spatially co-located macroblocks of previously correctly decoded frames. If there is not much motion, the replacement macroblocks will be quite similar to the ones that should have been decoded and the error will hardly be noticeable. In areas of faster motion, however, this concealment algorithm is less successful.

More sophisticated concealment algorithms use correctly decoded motion vectors of surrounding macroblocks in an attempt to motion compensate the missing areas from adjacent frames [9]. For this purpose MPEG-2 has a special option to transmit concealment motion vectors even on intra-coded macro-blocks. However, this feature is rarely used in decoders and the concept of concealment motion vectors has been dropped in MPEG-4 (Visual and AVC).

Compared to MPEG-2, the effect of bit errors in MPEG-4 (AVC) Main and High Profile tends to be more pronounced. This is due to the fact that most MPEG-4 (AVC) encoders use single slices for an entire picture in order to save bit rate, whereas MPEG-2 requires separate slice codes on every row of macroblocks. Furthermore, MPEG-4 (AVC) tends to use longer GOPs, thus increasing the time it takes before the bit error distortion is cleared.

The worst-case scenario, however, is if a bit error occurs in a picture or sequence header in MPEG-2 or in a picture or sequence parameter set in MPEG-4 (AVC). The likelihood of this happening is very low. In fact, it is several orders of magnitude smaller than the likelihood of bit errors occurring elsewhere in the bit stream. However, the effect of bit errors in headers can be considerably worse than that caused by errors elsewhere.

11.5   Post-processing

Video compression picture quality is largely determined by the algorithms used in the encoder. The decoder merely follows the instructions of the bit stream generated by the encoder. However, this does not mean that the decoder has no effect on the picture quality and the viewing experience in homes. In analogy with the pre-processing functions in an encoder, both optional and mandatory post-processing operations are carried out within the decoder.

Neither the 4:2:0 to 4:2:2 conversion filters, nor the horizontal up-sampling filters are defined in MPEG, although both are necessary operations in many DTH transmission systems. As we have seen in Chapter 8, filter design can have a significant effect on picture quality. This applies to both pre-processing and post-processing filters. In fact, the divergence in frequency response between different polyphase filters is even greater in up-conversion filters used in decoders than in down-conversion filters used in encoders. This can be seen by comparing the frequency response of up-conversion polyphase filters shown in Figure 11.4 below with those of the corresponding down-conversion filters shown in Figure 8.2.

In addition to the vertical chrominance up-conversion filter and the hori-zontal image up-sampler shown in Figure 11.1, large 'HDTV ready' flat-screen displays with integrated SDTV decoders require sophisticated de-interlacing and up-conversion processes to produce a high-quality SDTV image on a HD display. Since up-conversion tends to magnify compression artefacts, some decoder/display manufacturers are now integrating de-blocking and mosquito filters into their designs.

Post-processing de-blocking filters are particularly relevant to MPEG-2 and MPEG-4 (Visual) decoders because these algorithms do not feature in-loop

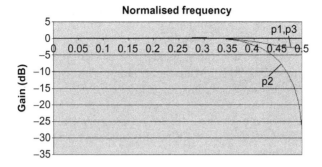

Figure 11.4   Frequency response of polyphase filters for a ⁴/₃ up-conversion

de-blocking filters as defined in MPEG-4 (AVC). Figure 11.5 shows how a post-processing de-blocking filter in a decoder can take advantage of parameters decoded from the bit stream. The de-blocking and/or mosquito filter can be controlled directly from the entropy decoder, which obtains the quantisation parameter (QP) for all macroblock positions from the bit stream. Using this information, the de-blocking filter can be controlled to filter only those picture areas that have been heavily quantised and leave undistorted picture areas unfiltered.

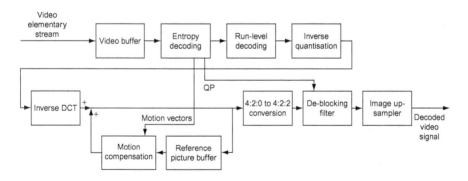

Figure 11.5   Block diagram of a video decoder with a de-blocking filter

11.6   Concluding remarks

Of course, there is much more to video decoders than just video and audio decoding. For a start, all digital set-top boxes feature an electronic programme guide (EPG) to provide additional programme information and easier search through a large number of channels. Furthermore, set-top boxes with return channels can provide interactive features such as information services, games, interactive voting, email, short message service (SMS) and home shopping by remote control. These applications are running on a middleware software. There are currently two open systems standards for interactive services: MHEG-5 (Multimedia and Hypermedia Experts Group, not to be confused with MPEG) and DVB-MHP (Multimedia Home Platform).

However, the most radical innovation in television viewing is probably the personal video recorder (PVR) [10]. By adding a hard disk to the decoder, it is possible to use trick modes such as pause, fast forward and rewind on live programmes [11]. Furthermore, PVRs using EPGs enable programme recordings with a single push of a button, thus making 'complicated' video cassette recorders (VCRs) obsolete.

11.7   Summary

- Channel change time between MPEG-4 (AVC) bit streams tends to be generally longer than between MPEG-2 bit streams due to larger buffer sizes and lower bit rates.
- Channel change time between different multiplexes is largely implementation dependent.
- Once a bit error has been detected by the decoder, the decoding process can only resume at the next slice header. In MPEG-2 each row of macroblocks starts with a slice header, whereas in MPEG-4 (AVC) slice headers are usually only at the start of a picture.

Exercise 11.1   Channel change time

A user changes channels on a satellite set-top box. Both channels are statistically multiplexed within the same TS. The GOP lengths of the bit streams varies between 12 and 15 frames (25 frames/s), the video buffer size is set to 1.2 Mbit and the bit rate varies between 1 and 10 Mbit/s. What is the minimum, maximum and average channel change time?

Exercise 11.2   Filter design

A filter bank of 16 polyphase filters has been designed for horizontal up-sampling. By choosing the polyphase filter with the closest phase position, the same filter bank can be used for any up-sampling ratio without changing the filter coefficients. Why is this not possible for down-sampling filters?

References

1. ETSI TS 101 154, 'Digital Video Broadcasting (DVB) Implementation Guidelines for the Use of Video and Audio Coding in Broadcasting Applications Based on the MPEG-2 Transport Stream', V1.5.1, 2005.

2. ATSC A/53 Digital Television Standard Part 4, 'MPEG-2 Video System Characteristics, with Amendment No. 1', 2007.

3. S.J. Huang, 'MPEG-2 Video Decoder Compliance Test Methodology', Canadian Conference on Electrical and Computer Engineering, September 1995. pp. 559–62.

4. International Standard ISO/IEC 13818-1 MPEG-2 Systems, 2nd edition, 2000.

5. T. De Couasnon, L. Danilenko, F.J. In der Smitten, U.E. Kraus, 'Results of the First Digital Terrestrial Television Broadcasting Field-Tests in Germany', *IEEE Transactions on Consumer Electronics*, Vol. 39, Issue 3, June 1993. pp. 668–75.

6. T.H. Tsai, Y.X. Lee, Y.F. Lin, 'Video Error Concealment Techniques Using Progressive Interpolation and Boundary Matching', International Symposium on Circuits and Systems, May 2004. pp. V/433–6.

7. P. Salama, N.B. Shroff, E.J. Delp, 'Error Concealment in MPEG Video Streams over ATM Networks', *IEEE Journal on Selected Areas in Communications*, Vol. 18, Issue 6, June 2000. pp. 1129–44.

8. D. Kim, J. Kim, J. Jeong, 'Temporal Error Concealment for H.264 Video Based on Adaptive Block-Size Pixel Replacement', *IEICE Transactions on Communications*, Vol. E89-B, 2006. pp. 2111–4.

9. S. Tsekeridou, I. Pitas, 'MPEG-2 Error Concealment Based on Block-Matching Principles', *IEEE Transactions on Circuits and Systems for Video Technology*, Vol. 10, Issue 4, June 2000. pp. 646–58.

10. S.-W. Jung, D.-H. Lee, 'Modeling and Analysis for Optimal PVR Implementation', *IEEE Transactions on Consumer Electronics*, Vol. 52, Issue 3, August 2006. pp. 77–78.

11. J.P. van Gassel, D.P. Kelly, O. Eerenberg, P.H.N. de With, 'MPEG-2 Compliant Trick Play over a Digital Interface', International Conference on Consumer Electronics, 2002. pp. 958–66.

Chapter 12
Statistical multiplexing

12.1 Introduction

Up to this point we have concentrated on the operation and performance of individual encoders and decoders. Although we have considered different operating points for mobile, SDTV and HDTV applications, we have always focussed on single encoders and decoders. In this chapter we will have a look at how several encoders can work together to improve the overall picture quality.

If MPEG encoders work in constant bit rate (CBR) mode, the picture quality varies depending on the criticality of the video signal. In order to generate a consistently high picture quality, the bit rate should change in accordance with the criticality of the video signal. A good example is MPEG-2 encoding on DVDs, where the bit rate changes from scene to scene from about 1 Mbit/s up to 10 Mbit/s, achieving an average bit rate of about 5 Mbit/s. A similar concept can be used in multi-channel transmission systems by sharing the total available bandwidth between a number of channels. Those encoders requiring higher bit rates can get a bigger share of the total bandwidth than those encoding non-critical material. Since the bit-rate variation depends on the statistics of the video signals, such systems are usually referred to as 'statistical multiplexing systems'.

MPEG signals demand high bit rates for short periods of times and much lower ones for most of the rest of the time. Picture quality, however, is usually judged on the worst parts of distortion. Therefore, CBR systems have to be set to a bit rate high enough so that even the most critical scenes are coded at an acceptable quality level. For most of the time, a lower bit rate would be perfectly adequate. If several video signals share the same transmission channel, more bit rates can be allocated to those channels that have the highest bit-rate demand for short periods of time by temporarily reducing the bit rate of the other channels. This evens out the peaks and troughs of the required bit rates and provides a net improvement on all channels [1,2].

Theoretically, there is a huge variation in the number of bits an MPEG encoder can generate for different video input signals. For example, the smallest full-resolution SDTV MPEG-2 (P) frame consists of only 344 bytes. This corresponds to a bit rate of 69 kbit/s and a compression ratio of 2 400. For MPEG-4 (AVC), the smallest SDTV frame size consists of a mere 34 bytes, corresponding to a bit rate of 6.8 kbit/s and a compression ratio of 24 400.

Conversely, the biggest full-resolution SDTV MPEG-2 frame can approach the bit rate of the uncompressed video signal including blanking. For example, a full-resolution SDTV frame, loaded with a maximum-level YUV noise and intra-coded with a *quantiser_scale_code* of 1 generates about 1.2 Mbytes, corresponding to a bit rate of 240 Mbit/s.

These variations in bit rate happen on different time scales. Table 12.1 summarises the factors that contribute to bit-rate variations on different time scales, from a few microseconds to several hours. Short-term variations due to macroblock coding types, picture detail and in particular picture coding types (I, P and B frames) can and should be absorbed in the rate buffer. Statistical multiplexing deals with longer-term variations, due to changes in scene criticality (pans, zooms, fades or scene cuts) or different types of programme material, such as sports, movies, news, etc. In other words, the encoder rate control evens out the bit-rate fluctuations of a video signal within a group of pictures (GOP), whereas statistical multiplexing deals with longer-term bit-rate demands between different encoders sharing a common group bit rate. Both types of algorithms are trying to keep picture quality constant.

Table 12.1   Bit-rate variations on different time scales

Entity	Time scale	Control mechanism
Macroblock coding type	10 µs	Rate control
Detail variation within a picture	< 40 ms	Rate control
Picture type (I, P or B picture)	40 ms	Rate control
Spatial activity (picture detail)	40 ms	Rate control
Temporal activity (pan, zoom, rotation, fade, etc.)	Several 100 ms to several seconds	Statistical multiplexing
Scene changes	Several seconds	Statistical multiplexing
Programme material (sports, news, movies, etc.)	Several hours	Statistical multiplexing
Channel type (sports channel, movie channel, etc.)	Permanent	Statistical multiplexing

Figure 12.1 shows a block diagram of a simple statistical multiplexing system. The encoders send the compressed variable-bit-rate video signals, as well as the current bit-rate demand, to the multiplexer. Instead of bit-rate demand, the encoders could send picture criticality, picture quality or some

other parameter related to the current bit-rate demand to the multiplexer. The multiplexer combines the video services into one CBR multi-programme bit stream, calculates new bit-rate allocations based on the current bit-rate demands of the encoders and sends the new allocated bit rates to the encoders. Note that the bit-rate allocation need not be calculated in the multiplexer itself. It could equally be calculated in a separate control computer.

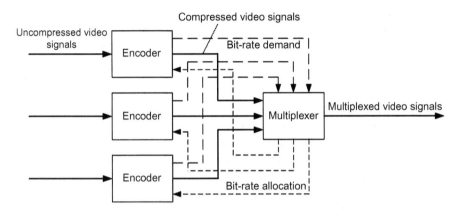

Figure 12.1   Block diagram of a simple statistical multiplexing system

12.2   Analysis of statistical multiplex systems

In order to analyse the performance of statistical multiplexing systems, we start by measuring the bit-rate demand of individual video channels. This can be done by setting an encoder into constant quality (variable bit rate) mode and measuring the bit-rate demand over a long period of time. Depending on the type of channel, the histogram of bit-rate demand will converge towards a stable probability density function (PDF) after several hours. Channels with different content converge towards different probability distributions of

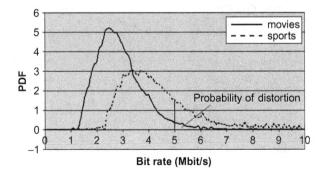

Figure 12.2   Histogram of bit-rate demand for two types of channels with constant quality

bit-rate demands, as shown in Figure 12.2. Once the probability curve has stabilised, the bit-rate characteristics of the channel can be analysed.

Since the PDF was generated with the encoder set to a given constant quality, we can use the PDF to work out probabilities of distortion. If we select a CBR of 5 Mbit/s for example, then the area of the PDF above 5 Mbit/s corresponds to the probability that the distortion is higher than that of the target quality with which the PDF was generated in the first place (since the total area under the PDF is 1).

In the example shown in Figure 12.2, the probability of the movie channel to require more than 5 Mbit/s is almost zero whereas that of the sport channel is close to 25 per cent. In other words, at 5 Mbit/s, the movie channel would exceed the required picture quality most of the time whereas the sports channel would suffer some distortion beyond the quality target for 25 per cent of the time. Therefore, the mechanism described at the beginning of this section can be used to calculate the required bit rate for a given quality target.

At this stage it is useful if we can simplify the analysis by developing a model for the bit-rate demand. A closer inspection of a number of bit-rate demand PDFs shows that while the lower bit rates can be approximated by a simple polynomial function, the higher bit rates tend to decay according to an exponential function.

Bit-rate PDF $\sim$ second-order polynomial for $R \leq R_{50}$

Bit-rate PDF $\sim$ e$^{(-a \times R)}$ for $R > R_{50}$

where a is a constant, R is the bit rate and R_{50} is the median bit-rate demand.

For a more detailed derivation and description of the model, see Appendix I and Reference 3. Figures 12.3 and 12.4 show that such a model closely matches the cumulative distribution function (CDF) and the PDF of the measured bit-rate demand, respectively. Using such a model, detailed calculations on the bit-rate saving of statistical multiplexing can be carried out.

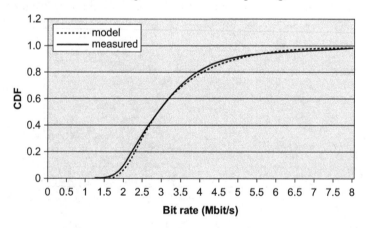

Figure 12.3   Model and measured cumulative distribution

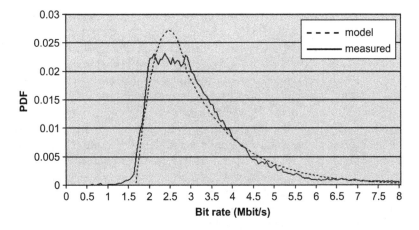

Figure 12.4   Model and measured probability density function

12.3   Bit-rate saving with statistical multiplexing

If two or more channels are statistically multiplexed, their combined bit-rate PDF can be calculated by convolution of the individual PDFs. The combined PDF converges towards a Gaussian distribution as more and more channels are statistically multiplexed, as shown in Figure 12.5. However, a Gaussian distribution drops off far more rapidly towards higher bit rates than an exponential function. This means that the probability of exceeding a certain bit-rate demand is smaller than with a single channel.

Gaussian bit-rate PDF $\sim e^{(-a \times R \times R)}$ for $R > R_{50}$

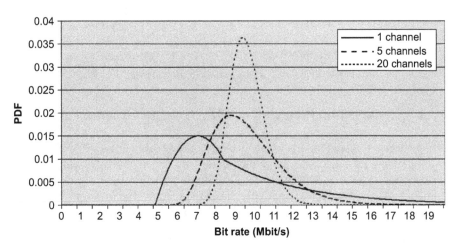

Figure 12.5   PDFs for 1, 5 and 20 channels statistically multiplexed

Figure 12.5 shows the PDF of the bit-rate demand (per channel) of 1, 5 and 20 statistically multiplexed HDTV channels. It can be seen that the required peak bit rate per channel drops as more and more channels are statistically multiplexed. At the same time, the average required bit rate (as well as the most likely required bit rate) increases with the number of channels. However, it is the peak bit rate, not the average, that determines the required total bit rate and the level of distortion at a given bit rate.

The bit-rate saving due to statistical multiplexing can be calculated by finding the bit rate at which the area underneath the PDF and to the right of the chosen bit rate is equal to, say, 1 per cent. One per cent means that during an average movie, the picture quality would drop below the required picture quality for less than 1 min. It can be seen in Figure 12.5 that the bit rate that is exceeded for 1 per cent of the time drops from 20 Mbit/s for a single channel to 16 Mbit/s for five channels and down to 13 Mbit/s for 20 channels. Figure 12.6 shows the bit-rate saving due to statistical multiplexing for statistical multiplexing groups of up to 15 channels. It can be seen that even a relatively small number of channels provides a significant bit-rate saving. Statistical multiplexing groups of more than 20 channels provide little additional bit-rate saving.

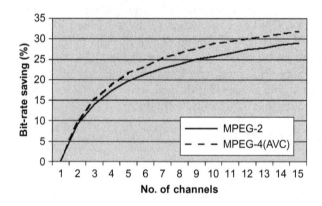

Figure 12.6   Bit-rate saving due to statistical multiplexing

12.4   Statistical multiplexing of MPEG-4 (AVC) bit streams

As has been explained in Chapter 5, there is a large number of new AVC coding tools capable of substantially reducing the bit-rate demand of non-critical sequences by reducing spatio-temporal redundancies. However, relatively few coding tools are as effective on highly critical sequences, the in-loop filter being one notable exception. This leads to two seemingly contradictory conclusions:

- Since a large percentage of broadcast material is non-critical, bit-rate savings in excess of 50 per cent might be expected.

- On the other hand, most viewers would judge the performance of compression systems by its worst artefacts, i.e. based on a small percentage of highly critical material. However, the bit-rate saving of AVC encoders on highly critical material is not quite as high as on non-critical sequences.

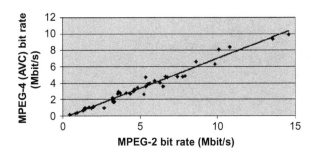

Figure 12.7   Scatter diagram of MPEG-4 (AVC) bit rates versus MPEG-2 bit rates for SDTV

This can be demonstrated by analysing an extreme example: The bit-rate demand of a white Gaussian noise sequence is approximately the same for MPEG-2 and MPEG-4 (AVC), assuming that the quality is set to a high level, so that the MPEG-4 loop filter is not active. Therefore, the real answer about bit-rate savings of MPEG-4 (AVC) against MPEG-2 can be derived only from a statistical comparison of the two systems. A first step is to compare bit-rate demands between MPEG-2 and MPEG-4 (AVC) for a given quality target.

12.4.1   Comparison of bit-rate demand of MPEG-2 and MPEG-4 (AVC)

Figure 12.7 shows a scatter diagram of MPEG-4 (AVC) bit-rate demand versus MPEG-2 bit-rate demand for a large number of SD test sequences at the same, relatively high picture quality. It can be seen that

1. the bit rates are surprisingly well correlated and
2. the percentage bit-rate saving is itself somewhat bit-rate dependent.

Note that the linear regression line does not cross the origin. Therefore, at low bit rates, the percentage bit-rate saving tends to be higher than at high bit rates. This point is illustrated more clearly in Figure 12.8, which shows a scatter diagram of MPEG-4 (AVC) bit-rate saving versus MPEG-2 bit-rate demand.

It can be seen that bit-rate savings of up to 70 per cent can be achieved with some non-critical sequences, whereas for the majority of sequences with medium-to-high criticality, the bit-rate saving is closer to 30 per cent.

Figure 12.9 also shows a scatter diagram of MPEG-4 (AVC) bit-rate demand versus MPEG-2 bit-rate demand, but this time for a set of HDTV test sequences at a picture format of 1 080 interlaced lines per frame and a frame rate of 25 frames/s. Again the bit-rate demand of the two algorithms is highly correlated, but this time the tendency of reduced bit-rate savings for higher criticality is even more pronounced, as shown in Figure 12.10.

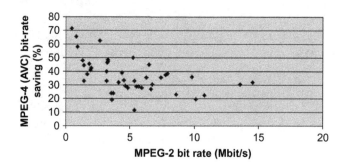

Figure 12.8   Scatter diagram of MPEG-4 (AVC) bit-rate saving versus MPEG-2 bit rates for SDTV

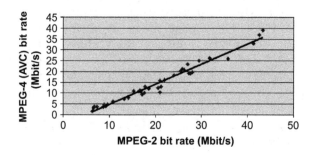

Figure 12.9   Scatter diagram of MPEG-4 (AVC) bit rates versus MPEG-2 bit rates for HDTV

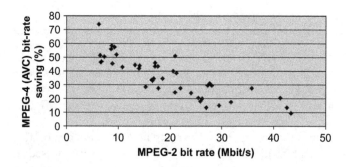

Figure 12.10   Scatter diagram of MPEG-4 (AVC) bit-rate saving versus MPEG-2 bit rates for HDTV

It can be seen that at a typical operating point (e.g. 15 Mbit/s MPEG-2 bit rate), the bit-rate saving is about 40 per cent, i.e. similar to that in SDTV. However, this reduces to less than 20 per cent for highly critical HDTV sequences, whereas in SDTV the average bit-rate saving remains 30 per cent, even for highly critical material. The reason for this somewhat surprising result is twofold.

- HDTV sequences tend to be more critical than SDTV sequences if the same camera techniques (pans and zooms, etc.) are used, because there is more pixel motion in HDTV than there would be in equivalent SDTV scenes.
- The comparison was carried out using a Main Profile encoder in both cases. The 4×4 transform of Main Profile encoders is somewhat less efficient in HDTV than it is in SDTV. Using a High Profile encoder with 8×8 transforms, the coding efficiency of MPEG-4 (AVC) HDTV encoders can be increased further.

The fact that bit-rate savings increase towards lower MPEG-2 bit rates (sequences with lower criticality) means that the total dynamic range of bit-rate demands is higher in MPEG-4 (AVC) than in MPEG-2. This makes statistical multiplexing even more effective in MPEG-4 (AVC) than in MPEG-2.

The linear correlation between the bit-rate demands of MPEG-4 (AVC) and MPEG-2 implies that the analysis applied to MPEG-2 statistical multiplexing systems outlined in Section 12.2 is equally applicable to MPEG-4 (AVC) SDTV and HDTV systems. For a more detailed investigation into statistical multiplexing of MPEG-4 (AVC), see Reference 4.

12.4.2 Bit-rate demand of MPEG-4 (AVC) HDTV

Having analysed the bit-rate demand of MPEG-4 (AVC) compression systems in relation to MPEG-2, it is possible to make bit-rate predictions of MPEG-4 (AVC) systems by calibrating the model parameters for MPEG-4 (AVC) encoders. For example, Figure 12.11 shows the PDF of the bit-rate demand of a single 1080i HDTV sports channel encoded in MPEG-4 (AVC) in comparison with MPEG-2, both encoded at the same picture quality. The picture quality is measured inside the encoders and held at a constant value, roughly equivalent to an ITU-R picture grade 4 (Good) [5] in both cases. Once the model parameters of a single channel have been determined, the bit-rate demand for statistically multiplexed systems can be derived by convolution of the single-channel PDFs.

The analysis in the previous paragraph has been carried out for a video format of 1 920 × 080 at 25 frames/s. A similar analysis can be carried out for a video format of 1 280 × 720 at 50 frames/s. Table 12.2 shows the results of such a comparison for constant picture quality. The comparison was carried out partially with sequences that are available in both formats ('New Mobile',

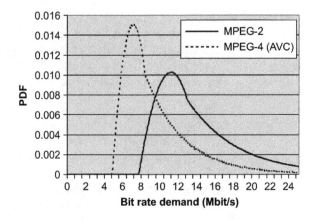

Figure 12.11   Comparison of bit-rate demand between MPEG-2 and MPEG-4 (AVC)

Table 12.2   Bit-rate comparison between 1080i25 and 720p50

Sequence	Bit-rate demand (Mbit/s)		Bit-rate saving (%) with 720p50
	1080i25	**720p50**	
New Mobile	9.6	5.6	41.8
Park Run	16.0	13.4	16.5
Stockholm Pan	5.3	4.7	11.6
Walking Couple	11.0	5.4	51.0
Vintage Car	7.2	4.1	42.5
Average	**9.8**	**6.6**	**32.7**

'Park Run' and 'Stockholm Pan') and partially with sequences that were down-converted from 1080i25 to 720p50 ('Walking Couple' and 'Vintage Car'). It can be seen that the bit-rate demand of 720p50 is approximately 30 per cent less than that of 1080i25.

12.5   System considerations

12.5.1   Configuration parameters

The new coding tools available in MPEG-4 (AVC) clearly have the most significant effect on picture quality improvements. However, there are a number of encoder configuration parameters that can also have considerable effects on picture quality. Obviously, the full benefit of statistical multiplexing can be utilised only if encoders (and decoders) support the full range of bit-rate demand as shown in Section 12.2. Furthermore, bit-rate changes must be adapted as quickly as possible on scene changes. If the actual bit-rate variation does not cover the full range of bit-rate demand (high as well as low bit rates), then the benefit of statistical multiplexing is compromised.

MPEG-4 (AVC) allows for bigger rate buffers than MPEG-2. This means that short peaks in bit-rate demand, e.g. during intra-coded frames, can be readily absorbed in the rate buffer without having to increase quantisation. Bigger rate buffers, however, imply longer end-to-end delays.

Unlike MPEG-2, where DCT and inverse DCT are defined only to a certain accuracy, the forward and inverse transforms in MPEG-4 (AVC) are defined bit accurate. Therefore, it is not necessary, from an encoding point of view, to insert intra-coded pictures at regular intervals in order to avoid decoder drift, as was the case in MPEG-2. In MPEG-4 (AVC), it would be sufficient to have I frames only at the beginning of each new scene. However, this would increase channel-change time, as pointed out in the previous chapter.

12.5.2 Noise reduction

Noise in video signals is by its very nature not predictable and therefore consumes a considerable amount of bits both in MPEG-2 and in MPEG-4 (AVC). Most video signals exhibit a certain amount of noise, even if it is hardly visible. The human eye is very efficient at discarding noise, so in many cases it may not be immediately noticeable. However, an encoder with a noisy input signal will demand a higher bit rate than one with a 'clean' signal (see Chapter 8).

Figure 12.12 shows how the effect of noise reduction can be modelled in terms of bit-rate demand. It can be seen that, although noise reduction has

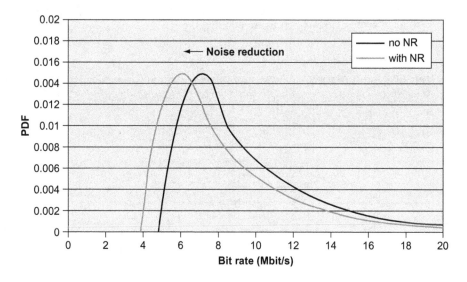

Figure 12.12   Effect of noise reduction on bit-rate demand

relatively little effect towards high bit rates, i.e. on critical material, it reduces the bit-rate demand for the majority of time on medium and lower criticality material. This frees up bit rate for other channels, while improving picture quality. Even small amounts of bit-rate reduction on every channel can add up to a significant improvement in picture quality on large statistical multiplex systems.

Most professional broadcast encoders have some form of integrated noise reduction [6]. Without noise reduction, encoders would waste a significant number of bits on encoding noise. Carefully controlled noise reduction helps to reduce the bit-rate demand of video signals, even if the noise level is almost invisible. This argument holds for both MPEG-2 and MPEG-4 (AVC) statistical multiplexing systems. For more on noise reduction, see Chapter 8.

12.5.3   Two-pass encoding

In conventional statistical multiplexing systems, the encoders report their bit-rate demands to the multiplexer, and the multiplexer then calculates and allocates new bit rates, which are sent back to the encoder. This involves a short delay between the time when the encoder realises that it needs a different (e.g. higher) bit rate and the time when it gets the new bit rate.

This problem can be overcome by using a look-ahead encoder that calculates the required bit rate ahead of time. The bit-rate demands are sent to the multiplexer before the video signal is actually coded. Before the video signal is compressed 'for real', it is delayed to allow for the turn-around time between bit-rate demands and bit-rate allocations. It is only when the new bit-rate allocations arrive back at the second encoder that the video signal from which the corresponding bit-rate demands are calculated is actually compressed. Figure 12.13 shows a block diagram of a two-pass encoder statistical multiplexing system. Two-pass encoding offers additional bit-rate savings, particularly in large systems of five or more encoders.

12.5.4   Opportunistic data

In statistical multiplexing systems, the bit-rate demand is calculated in each encoder. If an encoder can achieve the desired picture quality, then there is no

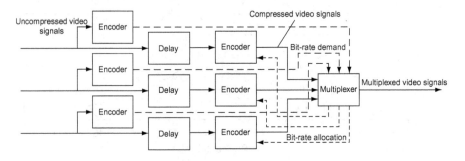

Figure 12.13   Block diagram of a statistical multiplex system with two-pass encoding

need to allocate a higher bit rate to it. Therefore, if a number of encoders are running at a relatively low bit rate because their video signals are non-critical, it is possible that there is some spare capacity in the transmission channel.

In conventional systems, this spare capacity will be either filled with null packets or distributed among the encoders. An alternative approach is to allocate the spare capacity to 'opportunistic' data services. This can increase the average data capacity of the data channel without degrading the picture quality of video signals.

12.6   Concluding remarks

Statistical multiplexing provides a coding benefit if three or more channels share the same transmission bandwidth. If the number of channels is quite low, it can improve picture quality by avoiding the worst artefacts during critical scenes. In larger statistical multiplexing groups, it can improve picture quality, as well as help to reduce the required total bit rate.

It is important to distinguish statistical multiplexing from variable bit rate (VBR) channel capacities. For example, IP networks could absorb high bit rates for short periods of time. However, whereas statistical multiplex systems are driven by the bit-rate demand of the source video signals, the channel capacity on IP network systems cannot be affected by the multiplex controller. As a result, the high bit-rate demand of a video signal might last longer than the channel capacity can absorb. Furthermore, the benefit of statistical multiplexing derives from the fact that the required bit rate is (generally) allocated as and when it is needed. This cannot be guaranteed in a variable bit-rate channel capacity. If a critical scene with high bit-rate demand coincides with network congestion, then compression artefacts get worse rather than better. Therefore, single-channel video streaming encoders are generally set to a CBR mode.

The situation is different when VBR encoding is used for storage media. Typical examples are encoding for DVDs or in PVRs. In these cases, the encoder can vary bit rate as and when required in order to keep the picture quality constant [7]. This is a form of statistical multiplexing in the time domain rather than across different video channels.

12.7   Summary

- CBR encoders have to be set to a bit rate close to the maximum bit-rate demand in order to avoid compression artefacts. Most of the time the encoder can cope with a lower bit rate.
- By sharing a common bit-rate capacity, encoders can free up bit rates to other channels during scenes of low criticality and obtain higher bit rates during critical scenes.
- By modelling the bit-rate demand for different content, the bit-rate saving of statistical multiplexing can be calculated.

- The bit-rate demand of MPEG-2 and MPEG-4 (AVC) encoders is strongly correlated although the bit-rate saving of MPEG-4 (AVC) increases for less critical scenes.
- Noise reduction frees up bit rate for other channels in the statistical multiplexing group.

Exercise 12.1   Statistical multiplexing at median bit-rate demand

A statistical multiplexing system consisting of five channels is configured to an extremely low bit rate such that the required picture quality is achieved for about 50 per cent of the time (median bit-rate demand). Can the picture quality be improved by adding more channels of similar criticality while keeping the average bit rate per channel the same?

Exercise 12.2   Large statistical multiplex system

A statistical multiplexing system has been expanded from 20 to 25 channels. Measuring the average picture quality of all channels it was found that the expansion provided little or no further improvement. Why did the expansion of the system not provide an improvement in picture quality or a further bit-rate saving?

References

1. M. Balakrishnan, R. Cohen, E. Fert, G. Keesman, 'Benefits of Statistical Multiplexing in Multi-Program Broadcasting', International Broadcasting Convention, Amsterdam, September 1997. pp. 560–5.
2. P.A. Sarginson, 'Dynamic Multiplexing of MPEG-2 Bitstreams', International Broadcasting Convention, Amsterdam, September 1997. pp. 566–71.
3. J. Jordan, A.M. Bock, 'Analysis, Modelling and Performance Prediction of Digital Video Statistical Multiplexing', International Broadcasting Convention, Amsterdam, September 1997. pp. 553–9.
4. A.M. Bock, 'Statistical Multiplexing of Advanced Video Coding Streams', International Broadcasting Convention, Amsterdam, September 2005.
5. ITU-R BT.500-11, 'Methodology for the Subjective Assessment of the Quality of Television Pictures', 2002.
6. A.M. Bock, 'What Factors Affect the Coding Performance of MPEG-2 Video Encoders', International Conference DVB'99, London, March 1999.
7. B.C. Song, K.W. Chun, 'A One-Pass Variable Bit-Rate Video Coding for Storage Media', *IEEE Transactions on Consumer Electronics*, Vol. 49, Issue 3, August 2003. pp. 689–92.

Chapter 13
Compression for contribution and distribution

13.1 Introduction

In the previous chapter we had a look at statistical multiplexing systems that are mainly used for large head-end transmission systems. However, MPEG compression is also used for point-to-point transmissions between broadcast studios and for distribution to local studios or transmitters.

The requirements for real-time video compression for contribution and distribution (C&D) links are often very different from those for direct-to-home (DTH) applications. Whereas DTH compression requires good picture quality at lowest possible bit rates, the priorities for C&D applications are more diverse. Typical contribution links are news feeds or feeds from sports events to studios or video links from one studio to another. Distribution links are from network centres to local studios or transmitter stations. In either case, the video signal often undergoes further processing after decompression, such as logo or ticker-tape insertion or chroma-key composition. Often the signal is re-encoded at lower bit rates for DTH transmissions. High picture quality and a minimum of compression artefacts are, therefore, critically important for such applications. Moreover, if chroma-key processing is applied after decoding, the spatial resolution of the chrominance signals is of utmost importance too. While the encode–decode end-to-end delay is less relevant for DTH applications, it can be an important factor for two-way live contribution links.

Prior to the availability of MPEG compression systems, high-bit-rate video links (34 Mbit/s for SDTV or more) were used for C&D applications [1]. Once it had been established that high picture qualities could be achieved with MPEG-2 at a fraction of the bit rates used at the time, it was clear that MPEG technologies could be used for C&D video links [2]. One hurdle to overcome was the introduction of a 4:2:2 Profile for Main and High Levels. Without the extended vertical chroma resolution of the 4:2:2 Profile, MPEG-2 could not have been used for video compression, which required chroma-key post-processing. Even if there is no further chrominance processing involved, the concatenation of several 4:2:0 down-sampling and up-sampling filters would lead to a rapid deterioration of the vertical chrominance bandwidth.

13.2   MPEG-2 4:2:2 Profile

13.2.1   Bit-rate requirement of 4:2:2 Profile

Although bit-rate demand might not be the highest priority for C&D systems, it is, nevertheless, important to know how much additional bit rate is required to encode a video signal in 4:2:2 format, as compared to 4:2:0. For this reason, a number of investigations have been carried out in order to measure the percentage of bits spent on chrominance for 4:2:2 and 4:2:0 coding [2,3]. In Reference 3, the measurements were carried out with the quantisation factor (QP) fixed to 3, which is typical for higher data rates in MPEG-2.

Table 13.1 gives a summary of the results. It can be seen that the percentage of data spent on chrominance varies quite significantly depending on content. In 4:2:2 coding, on average about twice as many bits are spent on chrominance than that in 4:2:0 coding as would be expected, but the amount does vary considerably. In general, camera material tends to have a higher level of chrominance saturation than film-originated content.

Table 13.1   Percentage of data spent on chrominance coding

Sequence	Percentage chroma in 422	Percentage chroma in 420
FTZM	42.79	21.43
Juggler	26.98	13.19
Mobile	28.81	14.74
Popple	23.72	14.09
Renata	22.69	9.52
Table Tennis	25.26	15.10
Tempete	45.28	24.58
XCL	19.53	6.46
Average	29.38	14.89

Since, under normal high-quality coding conditions, about 15 per cent of the data rate is allocated to chrominance information, the data rate required for 4:2:2 coding is about 15 per cent higher than that of 4:2:0 coding for a similar coding performance.

13.2.2   Optimum GOP structure with high-bit-rate 4:2:2 encoding

In 625-line MPEG-2 satellite broadcast systems, a GOP length of 12 frames is typically used either with or without B frames. Such a GOP length represents a useful compromise between coding efficiency and channel change time. At a bit rate of 20 Mbit/s, shorter GOP lengths still provide excellent picture quality [4], but may not be as efficient as longer ones.

Experiments on MPEG-2 4:2:2 Profile encoders have been carried out to measure the coding advantage of long GOPs versus short GOPs [3]. In these investigations the coding conditions were carefully controlled by making sure that the GOP structure was the only variable between coding runs. The tests were carried out at 20 Mbit/s, which is a typical bit rate for MPEG-2 contribution codecs. The results are shown in Table 13.2. It can be seen that the coding advantage of long GOPs is very significant even at a bit rate of 20 Mbit/s in SD.

Table 13.2   Signal-to-noise ratio at different GOP structures

Sequence	SNR dB $N = M = 2$	SNR dB $N = 12, M = 2$	SNR dB $N = 12, M = 1$	SNR dB $N = 2, M = 1$
Flower Garden	36.06	38.27	39.36	36.37
FTZM	47.61	52.35	52.32	47.54
Juggler	35.67	35.87	37.74	35.54
Mobile and Calendar	34.36	37.18	37.74	34.51
Popple	37.48	37.62	37.76	37.39
Renata	37.69	39.05	39.10	37.72
Susie	46.59	46.18	46.31	46.53
Table Tennis	37.98	37.98	38.74	38.06
Tempete	40.04	40.61	40.73	39.80
XCL	37.38	38.23	38.30	37.14
Average	39.09	40.33	40.81	39.06

Note: N = GOP length; M = distance between reference frames

The tests also showed that B frames offer no real coding advantage over purely forward predictive coding at high bit rates. Since B frames also increase the coding delay by at least two frame periods, it would be advantageous to disable the use of B frames in high-bit-rate, low-delay codecs.

As the bit rate is reduced to, say, 8 Mbit/s, the coding advantage of long GOPs over short GOPs increases and B frames start to become more efficient. Therefore, there is a trade-off between bit-rate efficiency and coding delay.

13.2.3   VBI handling in MPEG-2 4:2:2 encoding

MPEG-2 4:2:2 Profile at Main Level compression allows the definition of a picture area that exceeds the 576 normal active video lines of a 625-line television signal by 32 lines. This makes it possible to code teletext and other vertical blanking information (VBI) as part of the compressed MPEG video bit stream [5]. The advantage of this approach is that the VBI signals need not be decoded and put into a separate data stream, but they can be directly encoded together with the video signal. However, coded teletext lines are noise-like signals that are difficult to compress. Therefore, it is important to know how much extra bit rate has to be allowed for the teletext lines and which coding mode should be used for the compression of the teletext lines.

The bit-rate demand of teletext signals can be measured by filling the entire video frame with a simulated teletext signal. The teletext signal can be generated as a pseudo-random binary sequence (PRBS) sequence sampled at 6.9275 MHz, filtered down to 5.2 MHz and sample-rate converted to 13.5 MHz. This noise-like signal can then be MPEG-coded at a fixed quantisation, either as intra-frames or as non-motion-compensated P frames. Motion compensation is neither necessary nor useful in this case, because in practice motion compensation between teletext lines and normal active video should be strictly avoided. At the decoder, the decoded video signal is sample-rate converted back to 6.9375 MHz and compared with the original PRBS sequence in order to measure bit error rate (BER) and eye height. Eye height is a measure of integrity of a digitally transmitted signal at the receiving end. At an eye height near zero percent (eye closure) bit errors are inevitable.

The optimum coding mode can be determined by experimenting with a number of GOP structures. Table 13.3 shows that the best results are achieved with intra-coding, which performs significantly better than IPP, IBBP and IBP coding. Note that the bit rates quoted in Table 13.3 are theoretical minimum values with almost complete eye closure. In practice, a slightly higher bit rate should be allocated to the teletext lines. Since the results shown in Table 13.3 are based on a full-frame 576-line teletext signal, the bit rate per teletext line is calculated by dividing the measured bit rate shown in the table by 576.

Table 13.3   *Coding modes for MPEG-2 compression of teletext lines*

GOP structure	Max. QP for 0 errors	Bit rate full frame teletext (Mbit/s)
$N = M = 1$	20	28.9
$N = 12, M = 1$	13	42.0
$N = 12, M = 2$	14	34.1
$N = 12, M = 3$	13	33.7

To give an estimate of the actual bit-rate demand and QP, the eye height of the received signal should be measured in relation to the transmitted bit rate. The results are shown in Table 13.4.

It can be seen that with intra-coding ($N = M = 1$) one frame of teletext can be coded with a minimum bit rate of approximately 58 Mbit/s (QP = 10). This corresponds to a bit rate of 100 kbit/s per teletext line. With QP = 10, an eye height of more than 35 per cent is achieved, which should be sufficient under normal operating conditions. With IPP coding, one line of teletext needs a bit rate of 114 kbit/s (QP = 8) to achieve the same eye height, and with IBBP coding it needs 97 kbit/s (QP = 8). Figure 13.1 gives a graphical representation of the eye height as a function of QP.

Table 13.4   Effect of bit rate on received eye height for different coding modes

QP	$N = M = 1$		$N = 12, M = 1$		$N = 12, M = 3$	
	Bit rate (Mbit/s)	**Eye height (%)**	**Bit rate (Mbit/s)**	**Eye height (%)**	**Bit rate (Mbit/s)**	**Eye height (%)**
1	149.064	47.58	145.132	46.36	135.934	46.36
2	119.837	46.36	117.914	45.14	106.246	45.14
3	104.278	46.36	100.682	43.92	89.142	43.92
4	91.911	43.92	89.371	42.7	78.345	42.7
5	84.936	43.92	80.883	40.26	70.538	40.26
6	78.186	42.7	74.603	36.6	64.581	39.04
7	74.202	42.7	69.622	37.82	59.796	34.16
8	69.601	40.26	65.734	35.38	55.898	35.38
9	63.105	39.04	58.85	30.5	49.164	29.28
10	57.895	35.38	53.539	25.62	44.099	24.4
11	53.557	35.38	49.011	18.3	39.928	17.08
12	49.847	31.72	45.453	17.08	36.69	6.1
13	46.712	29.28	42.006	9.76	33.715	1.22
14	44.009	28.06	39.217	−6.1	31.321	−1.22
15	41.63	24.4	36.666	−3.66	29.176	−13.42
16	39.535	24.4	34.594	−6.1	27.411	−25.62
17	36.019	17.08	30.8	−23.18	24.275	−47.58
18	33.212	17.08	27.868	−34.16	21.878	−64.66
19	30.87	7.32	25.237	−53.68	19.774	−76.86
20	28.895	3.66	23.136	−67.1	18.113	−82.96
21	27.214	4.88	21.251	−67.1	13.042	−87.84

As has been shown above, direct MPEG compression of teletext signals requires a significantly higher bit rate than the teletext signal itself would need. Although, in principle, the same concept could be applied to MPEG-4 (AVC), the additional bit rate required for the teletext signal would, to some extent, negate the bit-rate saving of the advanced compression algorithm.

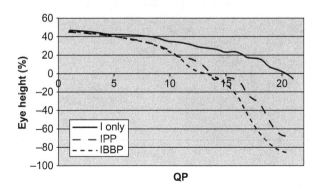

Figure 13.1   Received eye height as a function of transmitted MPEG-2 QP

13.3   MPEG-4 (AVC)

Initially, the main driving force for the rapid penetration of MPEG-4 technology into more and more HDTV and SDTV broadcast and IPTV applications was the high compression efficiency of MPEG-4 (AVC). Therefore, MPEG-4 (AVC) encoders have, up to now, mainly been optimised for bit rates substantially lower than MPEG-2 broadcast encoders. While MPEG-2 encoders might have been optimised for 3 Mbit/s full resolution SD, current MPEG-4 (AVC) encoders are likely to work best at 1.5 Mbit/s and full resolution. Similarly, HDTV MPEG-2 encoders need approximately 15 Mbit/s for the compression of 1080 interlaced video, whereas MPEG-4 (AVC) encoders would produce an equivalent picture quality at less than 7 Mbit/s. A more detailed comparison between MPEG-2 and MPEG-4 (AVC) at low bit rates was carried out some time ago [6]. However, one should bear in mind that estimates of equivalent constant bit rates can only give a rough indication of compression performance. A better understanding of bit-rate savings can be obtained by comparing statistical bit-rate demands of MPEG-4 (AVC) and MPEG-2 encoders over a range of different types of material [7].

Figure 13.2 shows the rate-distortion graphs for MPEG-2 and MPEG-4 (AVC) for a typical SDTV sports sequence (Soccer). This graph can be interpreted in two ways:

- The improvement in peak signal-to-noise ratio (PSNR) of MPEG-4 over MPEG-2 is almost constant over a wide range of medium-to-high bit rates.
- The percentage bit-rate saving of MPEG-4 compared to MPEG-2 increases sharply towards lower bit rates.

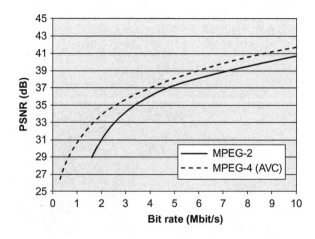

Figure 13.2   Typical sequence coded in MPEG-2 and MPEG-4 (AVC) SDTV

The latter point is made clearer in Figure 13.3, where the percentage bit-rate saving is plotted against MPEG-2 bit rate. Although the bit-rate saving of MPEG-4 versus MPEG-2 reduces with higher bit rates, significant bit-rate savings are still possible even at higher bit rates, assuming that both types of encoders have been optimised for such operating points.

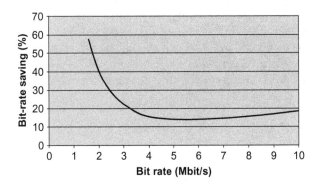

Figure 13.3   Bit-rate saving of MPEG-4 (AVC) SDTV at low bit rates

Interestingly, as the bit rate is increased even further, the bit-rate saving of MPEG-4 increases again, as shown in Figure 13.4. It is important to note that this is not due to encoder optimisation for a particular operating point, but a result of the fact that the gain in PSNR is constant and the gradient of the rate-distortion curve reduces towards higher bit rates. Therefore, constant PSNR gains translate to higher percentage bit-rate savings at higher bit rates. For these reasons it is expected that advanced video compression algorithms will find their way into C&D applications [8].

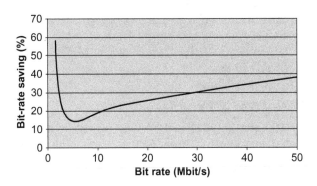

Figure 13.4   Bit-rate saving of MPEG-4 (AVC) SDTV at high bit rates

13.3.1 CABAC at high bit rates

One potential problem with running MPEG-4 (AVC) at higher bit rates is the real-time operation of CABAC [9]. While implementation of real-time CABAC encoding at low bit rates and with relatively large buffers is not too difficult, the bursty nature of the CABAC bit-stream output at low quantisation values makes its real-time implementation at high bit rates more difficult [10]. Therefore, the use of CAVLC or adaptive switching between CAVLC and CABAC may be considered for high bit rate MPEG-4 (AVC) systems. Fortunately, the coding gain of CABAC compared to CAVLC is at its highest at very low bit rates, where coding gains of up to 18 per cent can be achieved [11]. At higher bit rates, the coding gain of CABAC drops to about 10 per cent, as can be seen in Figure 13.5.

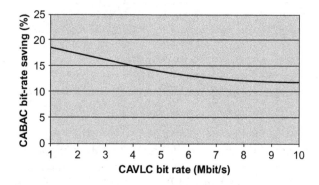

Figure 13.5   Bit-rate saving of CABAC versus CAVLC

However, high compression performance is not the only reason why MPEG-4 might find its way into C&D applications. A potentially more important factor is the fact that MPEG-4 (AVC) defines profiles that not only allow 4:2:2 transmission at 8 bit accuracy (as does MPEG-2) but also at 10 and even up to 14 bit accuracy. Since 10 bit 4:2:2 is the most common video standard in studio applications, the corresponding MPEG-4 (AVC) profile is ideally suited for C&D links.

13.3.2 MPEG-4 (AVC) profiles

As outlined in Chapter 5, MPEG-4 (AVC) defines a number of profiles applicable to video conferencing, broadcast and streaming applications:

- The Baseline Profile is mainly intended for video conferencing and streaming to mobile devices. It does not support B frames, interlace coding tools or CABAC.
- The Main Profile allows bi-directionally predicted (B) frames with two direct modes: spatial and temporal and weighted predictions. Furthermore,

it supports all interlace coding tools including picture adaptive field/frame coding (PAFF) and macro-block adaptive field/frame coding (MBAFF) as well as CABAC.

• The coding tools of MPEG-4 profiles that go beyond Main Profile are summarised as Fidelity Range Extensions [12]. In particular, the High Profile allows adaptive 8×8 integer transforms, intra 8×8 predictions modes and scaling lists.

• The High 10 Profile allows coding of 4:2:0 video signals with 10 bit accuracy and

• The High 4:2:2 Profile allows coding of 4:2:2 video signals with 10 or 8 bit accuracy.

• There is also a High 4:4:4 Profile for studio applications, which is not considered here because it is rarely used in C&D applications.

13.3.3 Fidelity range extension

MPEG-4 (AVC) High Profile provides a number of additional coding tools, which are particularly useful for HDTV compression. The most important of these is probably an integer 8×8 transform, which can be chosen in areas of less detail as an alternative to the conventional 4×4 transform. Furthermore, it defines a total of nine 8×8 intra-prediction modes, which are available in addition to the nine 4×4 and three 16×16 intra-prediction modes of the Main Profile. Last but not least, the High Profile allows user-definable scaling lists, which are the equivalent of MPEG-2 quantisation matrices.

Figure 13.6 shows a comparison of Main Profile and High Profile rate-distortion curves for 1080 interlaced HDTV sports content. It can be seen that at extremely low bit rates the differences are small, but at medium and higher bit rates a bit-rate saving of about 10 per cent can be achieved. Further tests have shown that bit-rate savings of about 9 per cent can be achieved for 720p HDTV material, whereas for SDTV the bit-rate saving is closer to 6 per cent.

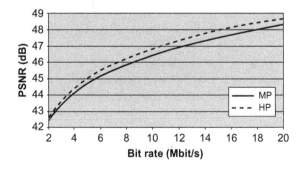

Figure 13.6 HDTV sports sequence coded in Main Profile and High Profile

13.3.4   4:2:2 MPEG-4 (AVC)

For contribution applications, 4:2:2 transmission is particularly relevant because it avoids repeated chroma down-sampling and up-sampling, which not only incurs a loss of vertical chrominance detail but also impedes further video processing. Furthermore, 10 bit 4:2:2 is the most commonly used studio format. From the point of view of compression efficiency, it is particularly important to know at what bit rate the picture quality of 10 bit 4:2:2 would be noticeably better than 8 bit 4:2:2.

The difference between 8 and 10 bit pixel precision is most noticeable in areas of low detail with subtle changes of colour. Most 8 bit source images contain enough noise to hide any contouring that might otherwise be visible in such areas. However, the inherent noise-reducing properties of video compression can expose the colour graduations and lead to a posterisation effect, even at relatively high bit rates.

Interestingly, there are two mechanisms in MPEG-2 that tend to hide such artefacts: DCT inaccuracy and mismatch control. Mismatch control is required as a result of DCT inaccuracy. The combined effect of these two mechanisms is to inject a small amount of noise at the output of MPEG-2 decoders. This noise acts like a low-level dither signal. Therefore, although MPEG-2, like most compression algorithms, eliminates some of the original source noise due to the low-pass characteristics of its weighting matrices, it reintroduces enough DCT noise at the decoder output to reduce posterisation.

This effect is illustrated in Figure 13.7. Figure 13.7a shows an extract of a plain area in an SDTV frame. In this picture the contrast has been enhanced by a factor of 10 in order to make the low-level noise visible. The SDTV sequence has then been encoded in MPEG-2 and MPEG-4 (AVC) at high bit rates in order to avoid noticeable compression artefacts. However, if we examine the decoded pictures in more detail, we find that there are differences between MPEG-2 and MPEG-4 (AVC). Figure 13.7b and c show the same extracts with contrast enhancement after MPEG-2 and MPEG-4 (AVC) decoding, respectively. It can be seen that whereas MPEG-2 adds high-frequency DCT noise, MPEG-4 (AVC) reproduces flat, noiseless areas, which make the contours between single grey levels differences (with 8 bit resolution) more visible.

The disadvantage of MPEG-2 DCT inaccuracies is, of course, decoder drift. This limits the number of predictions that can be made from previous predictions before a refresh is required. In other words, it limits how many P frames can be coded before another intra-coded frame has to be inserted to eliminate decoder drift. MPEG-4 (AVC) does not suffer from DCT inaccuracies because its integer transformations have no rounding differences between encoder and decoder. Therefore, it has no need for mismatch control and there is no limit to the number of predictions that can be made from previous predictions.

Theoretically an entire movie could be encoded with a single I frame at the beginning of the film. In practice, however, this is seldom the case because it

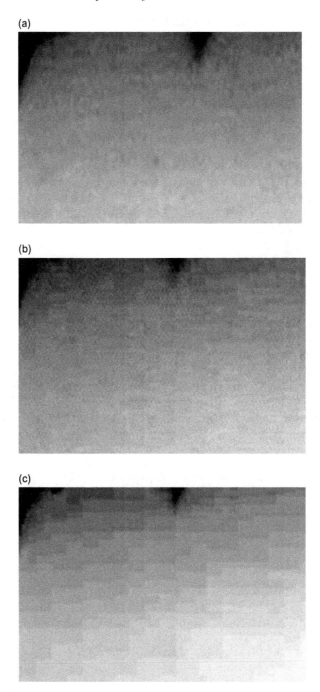

*Figure 13.7   (a) Low detail area of a non-critical sequence. Extract of
                (b) MPEG-2 and (c) MPEG-4 (AVC) decoded picture area.*

would make it difficult to select individual sections of the film. A more appropriate application is in low-delay point-to-point links. Since I frames require large video buffers, the end-to-end delay can be reduced by avoiding I frames altogether. This is possible because the decoder can gradually build up the picture if there are enough intra-coded macroblocks present.

Although the noise-reducing properties are less in MPEG-4 (AVC) than in MPEG-2 (due to its smaller 4×4 transform with unweighted quantisation), once source noise is removed, colour contouring, if present, remains exposed, unless artificial dither noise is injected at the decoder output. Adding a small amount of dither noise might be a good solution for consumer devices but is unlikely to be a satisfactory solution in professional decoders because it increases concatenation loss. As a result, the need for 10 bit coding could become apparent more quickly in MPEG-4 (AVC) systems than it did in MPEG-2.

13.3.5   Comparison of 8 and 10 bit coding

An important question to answer when comparing 8 with 10 bit video signals is to decide which assessment method should be used for the comparison. Since we are dealing with relatively high picture qualities, objective methods such as structural similarity (SSIM) [13] and PSNR are preferable to subjective assessment methods. The latter tend to become less accurate with higher picture quality because distortions are less obvious. Indeed, a comparison of SSIM with PSNR shows that the two are highly correlated as can be seen in Figure 13.8. Therefore, at very high bit rates, picture quality can be measured by carefully designed PSNR measurements (see also Chapter 3 on video quality measurements).

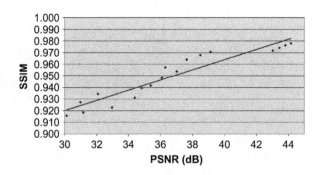

Figure 13.8   Scatter diagram of SSIM versus PSNR

In order to compare YCbCr PSNR results for 4:2:0 with those for 4:2:2, a method to combine individual PSNR figures for Y, Cb and Cr into a single number has to be defined. Since chrominance distortion is far less visible than luma distortion, a weighted sum of $0.8Y + 0.1Cr + 0.1Cb$ is used. These weighting factors have been proposed in the past [14]. However, it is a relatively arbitrary definition, which has not been verified by extensive subjective tests. Therefore, the comparison between 4:2:0 and 4:2:2 has to be interpreted with some caution. Nevertheless, such a definition allows us to investigate the trade-off between chroma resolution and picture distortion in different test sequences.

Figure 13.9 shows example PSNR curves for three SDTV MPEG-4 (AVC) compression formats at relatively low bit rates and average criticality. The three formats are: 8 bit 4:2:0, 8 bit 4:2:2 and 10 bit 4:2:2. The 10 bit 4:2:0 format was not included in these tests because it is assumed that 4:2:2 coding will be considered more important for C&D applications than 10 bit coding. While there is a noticeable difference between 4:2:0 and 4:2:2 at bit rates above 4 Mbit/s, the difference between 8 and 10 bit coding is insignificant at bit rates below 10 Mbit/s.

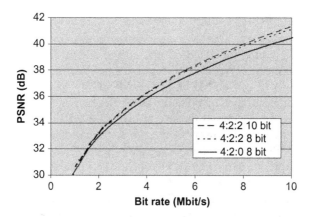

Figure 13.9   Comparison of three compression formats at relatively low bit rates

At higher bit rates, in this particular case at bit rates above 20 Mbit/s, the difference between 8 and 10 bit coding is becoming more noticeable, as can be seen in Figure 13.10. However, one has to bear in mind that these comparisons are highly sequence dependent. The graphs shown in Figures 13.9 and 13.10 are based on an SDTV sequence with average criticality. In non-critical sequences such as the extract shown in Figure 13.7a, the difference between 8 and 10 bit would be evident at much lower bit rates. Interestingly, there is no cross-over point between 8 and 10 bit coding, i.e. 10 bit video codes as efficiently as 8 bit video at low bit rates. Similarly, the cross-over point between 4:2:0 and 4:2:2 is at very low bit rates, as has been observed in MPEG-2 systems [15].

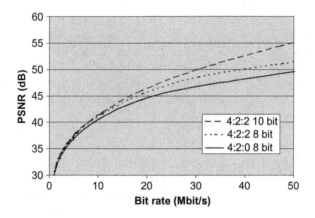

Figure 13.10   Comparison of three compression formats at relatively high bit rates

13.3.6   Low-delay MPEG-4 (AVC)

Point-to-point links used for live interviews require low end-to-end delay codecs. In this chapter the issue of low end-to-end delay has already been mentioned several times. At this point it might be constructive to compare the topic of low latency between MPEG-2 and MPEG-4 (AVC) algorithms.

According to ISO/IEC 14496-10, the rate buffer of MPEG-4 (AVC) MP@L3 can be up to 5.5 times the size of MPEG-2 MP@ML. Since MPEG-4 (AVC) is typically used at lower bit rates, end-to-end delays tend to be much higher than in MPEG-2. However, apart from hierarchical B frame structures, most of the compression tools of MPEG-4 (AVC) can be implemented without a significant increase in the latency of the encoder or the decoder, compared to the delay through the rate buffer. Since the ratio of the size of intra-coded frames relative to predicted frames is no higher in MPEG-4 (AVC) than it is in MPEG-2, it is possible to design low-latency MPEG-4 (AVC) encoders with higher performance than in MPEG-2. The only caveat in this argument is the potential challenge of using CABAC at high bit rates with small rate buffers as explained above. Indeed, the higher efficiency of I and P frame coding makes it possible to achieve higher efficiency at lower end-to-end delay. Of course, there is still a trade-off between delay and coding efficiency, as in any compression standard.

13.4   Concluding remarks

The substantial bit-rate savings of MPEG-4 (AVC) over MPEG-2 explains the rapid deployment of MPEG-4 (AVC) systems, in particular for new systems such as DTH HDTV and IPTV. At higher bit rates, typically used for C&D applications, the bit-rate savings are less obvious, particularly if end-to-end delay becomes a significant factor. Although the coding benefits of MPEG-4 (AVC) will eventually replace MPEG-2 across all applications, it is the 10 bit 4:2:2 Profile that has a unique advantage over all other compression formats.

13.5 Summary

- The rate-distortion curves indicate that the percentage bit-rate savings of MPEG-4(AVC)versusMPEG-2ishighestatverylowandveryhighbitrates.
- The bursty output of CABAC encoding at low quantisation makes it less suitable for high-bit-rate, low-delay applications.
- MPEG-2 DCT inaccuracies and mismatch control acts like a dither signal, thus reducing posterisation.
- The MPEG-4 (AVC) High 10 Profile avoids posterisation in plain areas without requiring a higher bit rate.
- The MPEG-4 (AVC) High 4:2:2 10 bit Profile is ideally suited for studio-to-studio contribution applications.

Exercise 13.1   Distribution link

A broadcast company is distributing three MPEG-4 (AVC) SDTV channels simultaneously to its local studios. Should they use a constant bit rate of 10 Mbit/s to protect critical sports material or would statistical multiplexing be beneficial?

References

1. D.I. Crawford, 'Contribution and Distribution Coding of Video Signals', International Broadcasting Convention, Amsterdam, September 1994. pp. 188–95.
2. C.H. Van Dusen, A. Tabatabai, B.J. Penney, T. Naveen, 'MPEG-2 4:2:2@ML From Concept to an Implementation', International Broadcasting Convention, Amsterdam, September 1996. pp. 386–90.
3. A.M. Bock, G.M. Drury, W.J. Hobson, 'Broadcast Applications of 4:2:2 MPEG', International Broadcasting Convention, Amsterdam, September 1997. pp. 123–28.
4. J.H. Wilkinson, 'The Optimal Use of I, P and B Frames in MPEG-2 Coding', International Broadcasting Convention, Amsterdam, September 1996. pp. 444–49.
5. O.S. Nielsen, F.L. Jacobsen, 'Multichannel MPEG-2 Distribution on Telco Networks', International Broadcasting Convention, Amsterdam, September 1996. pp. 601–5.
6. J. Bennett, A.M. Bock, 'In-depth Review of Advanced Coding Technologies for Low Bit Rate Broadcast Applications', *SMPTE Motion Imaging Journal*, Vol. 113, December 2004. pp. 413–18.

7. A.M. Bock, 'Statistical Multiplexing of Advanced Video Coding Streams', International Broadcasting Convention, Amsterdam, September 2005. pp. 193–200.

8. Y. Zheng, A. Lipscombe, C. Chambers, 'Video Contribution over Wired and Wireless IP Network – Challenges and Solutions', BBC Research White Paper, WHP 150, July 2007.

9. A.M. Bock, '4:2:2 Video Contribution and Distribution in MPEG-4 AVC', NAB Broadcast Engineering Conference, Las Vegas, April 2006. pp. 417–21.

10. L. Li, Y. Song, T. Ikenaga, S. Goto, 'A CABAC Encoding Core with Dynamic Pipeline for H.264/AVC Main Profile', IEEE Asia Pacific Conference on Circuits and Systems, December 2006. pp. 760–63.

11. D. Marpe, H. Schwarz, T. Wiegand, 'Context-Based Adaptive Binary Arithmetic Coding in the H.264/AVC Video Compression Standard', *IEEE Transactions on Circuits and Systems for Video Technology*, Vol. 13, Issue 7, July 2003. pp. 620–36.

12. D. Marpe, T. Wiegand, S. Gordon, 'H.264/MPEG-4 AVC Fidelity Range Extensions: Tools, Profiles, Performance, and Application Areas', Proceedings of ICIP, Genoa, September 2005. pp. 593–96.

13. Z. Wang, A.C. Bovik, H.R. Sheikh, E.P. Simoncelli, 'Image Quality Assessment: From Error Visibility to Structural Similarity', *IEEE Transactions on Image Processing*, Vol. 13, No. 4, April 2004. pp. 600–12.

14. Z. Wang, L. Lu, A.C. Bovik, 'Video Quality Assessment Based on Structural Distortion Measurement', *Signal Processing: Image Communication*, Vol. 19, February 2004. pp. 121–32.

15. A. Caruso, L. Cheveau, B.G. Flowers, 'The Use of MPEG-2 4:2:2 Profile@ML for Contribution, Collection and Primary Distribution', International Broadcasting Convention, Amsterdam, September 1998. pp. 539–44.

Chapter 14

Concatenation and transcoding

14.1 Introduction

In the previous two chapters, we have seen that MPEG video compression is used for contribution and distribution (C&D) applications as well as for statistical multiplexing systems in DTH head-end systems. Today, it is inevitable that video signals undergo several stages of compression and decompression before it reaches the end user, and with the growth of digital television networks, concatenation of compression encoding and decoding is becoming more and more prevalent. In this chapter we will have a closer look at all combinations of concatenation between MPEG-2 and MPEG-4 (AVC).

In practice, there are different scenarios of compression concatenation, ranging from a closely coupled decoder followed by an encoder to a distributed system where the decoder is completely independent of the encoder and there could be several processing stages in between. The former is usually described as a transcoder, whereas the latter is referred to as a concatenated compression system. In this chapter we concentrate on concatenated systems although many aspects of this chapter are also applicable to loosely coupled transcoders. A more detailed description of transcoders is provided in Chapter 15.

Some years ago, a number of proposals were published to improve the cascade performance of video compression systems [1,2]. The common feature of these proposals was the requirement to convey previous compression parameters to downstream encoders through a side channel, which is carried, together with the video signal itself, from the decoder to the following encoder. In particular, the use of the two least significant chrominance bits of a 10 bit digital video signal has been proposed as a means of carrying helper information for re-encoding [3]. Examples of compression parameters are the picture coding type (i.e. whether a picture is intra-coded, predictively coded from previous frames or bi-directionally predicted from past and future frames), quantisation levels, motion vectors, mode decisions, etc. While such systems have been demonstrated to work well under controlled conditions, the integrity of the side channel cannot always be guaranteed in practice.

A second disadvantage of a system with a side channel is the data capacity required for such a helper signal. In fact, for the full re-coding data set of an MPEG-2 compressed video signal, a channel capacity of approximately

20 Mbit/s is required [4]. However, the importance of the compression para-meters in terms of their effect on the picture quality in subsequent compression stages varies greatly. In particular, the picture coding type of previously com-pressed video signals is significantly more important than all the other para-meters put together. This information, however, could be carried in less than 50 bit/s.

However, the real reason why today's systems do not utilise helper infor-mation in concatenated encoders is the cost and therefore the lack of avail-ability of MPEG decoders that can provide such metadata information in the first place. While it would be relatively simple to extract some compression parameters from MPEG headers and carry them on a side channel, other parameters, such as quantisation and motion vectors, require full variable-length decoding and can be made available only in special decoder chips or FPGAs that have been designed for that purpose. Furthermore, pixel syn-chronisation of video signal and side channel may present an implementation problem in practice. In any case, integrated MPEG-2 or MPEG-4 (AVC) decoder chips that provide side channel information are still not commercially available, more than 10 years after the original proposal.

In the first part of this chapter, we will examine the effects of repeated MPEG compression/decompression on picture quality from first principles and present a model describing the accumulation of compression noise over several generations of generic MPEG encoding and decoding. Furthermore, this chapter describes how the knowledge of the statistical properties of MPEG compression noise can be used to derive important compression parameters from the video signal itself, thus avoiding the need for a helper channel altogether.

14.2   Concatenation model

In order to analyse the effects of MPEG compression parameters on the per-formance of concatenated encoders, a large number of distortion measure-ments on a chain of two MPEG-2 encoder/decoder pairs have to be carried out. Figure 14.1 shows the set-up for such measurements. The distortion can be measured both visually and in terms of PSNR values. In performing such tests, it has been found that the picture quality of cascaded encoders depends on a number of factors, such as encoder configurations, picture material, bit rates, the order of encoding, GOP alignment, etc. [5]. Furthermore, it was found that while in some cases the picture quality of two cascaded encoders was similar to that of a single encoder running at the same bit rate, in other cases, con-catenation of two encoders severely degraded the picture. A method of describing such diverse results in a single model is described in the following paragraph.

The proposed model splits the compression noise N1 and N2 of the two encoders into two components: one that is in phase to the compression noise of

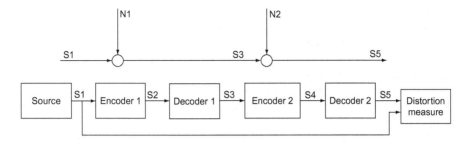

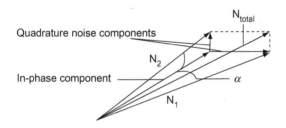

Figure 14.1   Measurement set-up and signal diagram

the previous encoder and one that is in quadrature to the noise of the previous stage and to the in-phase component. Figure 14.2 gives a graphic representation of the noise components in the case of two cascaded encoders. The model can be extended to more than two compression stages, but the parameters become more difficult to quantify. Note that the noise components in Figure 14.2 represent the total compression noise generated by encoders, i.e. quantisation noise as well as other noise components that are due to motion compensation, etc.

Figure 14.2   Model of concatenation noise

The purpose of this model is to illustrate the effect of the compression noise of concatenated encoders. While each encoder generates its own compression noise, a certain portion of this noise is identical to the noise generated in the previous encoder. This portion of the noise is labelled the 'in phase component'. The higher the in-phase component in comparison to the quadrature noise components, the higher the end-to-end picture quality of the system.

Based on this model, MPEG cascade performance can be expressed in terms of the noise correlation angle α. A correlation angle of $\alpha = 0$ implies that the second encoder completely (re-)generates the noise of the first encoder. Conversely, if $\alpha = 90°$, the two noise components are completely uncorrelated and the total noise power is the sum of $N_1^2 + N_2^2$. The noise correlation depends on video content, bit rate, encoder configuration and GOP alignment. The higher the bit rate and the lower the coding criticality, the higher the correlation between the two noise components. However, in order to reduce the correlation angle close to zero, we also have to use identical encoder configurations and GOP alignment between the two encoders. On the other hand, any processing carried out on S3, such as composite encoding and decoding, noise reduction, filtering, etc., is likely to destroy the correlation between the two noise components and thus reduce the picture quality of the overall system, even if the picture quality of the intermediate signal, S3, is slightly improved.

Figure 14.3 gives a graphical representation of the measured results over up to six encode/decode stages. The measurements shown in Figures 14.3 and 14.4 were carried out with test sequences of average criticality. Less critical sequences would produce a smaller drop in PSNR, whereas highly critical sequences would produce a larger drop.

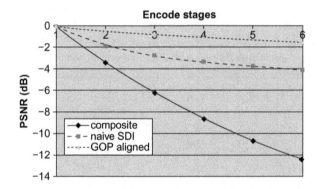

Figure 14.3   Measured results over several generations of MPEG-2 concatenation

It can be seen in Figure 14.3 that composite encoding and decoding between successive compression stages leads to a rapid deterioration of picture quality. Serial digital links between decoders and encoders improve the concatenation performance considerably, but there is still a quality loss from one encoder to the next. This is because frames might be coded as B frames in one encoder and as I or P frames in another. The best performance can be achieved if all encoders are set up to the same GOP length and all GOP structures are aligned.

Figure 14.4 shows how the picture quality varies as the I frame of the second encoder is shifted with respect to the I frame position of the first encoder. It can be seen that the highest quality (highest PSNR) is achieved if I frames are re-encoded as I frames, B frames as B frames and P frames as P frames. If the I frame position is shifted by three, six or nine frames, the picture quality is higher than at the remaining offset positions because in those cases the I frame of the second encoder is aligned with a P frame of the first encoder.

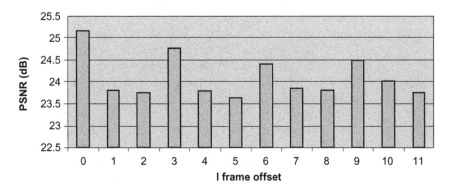

Figure 14.4   Picture quality with different GOP alignment (position 0 is I frame aligned)

Since the quantisation noise of P frames is slightly lower than of B frames, this gives a better end-to-end picture quality.

If all encoders are I frame synchronised, set up identically and operating at the same bit rate, cascading loss can, theoretically, be reduced further, almost down to zero, although this is difficult to achieve in practice. In most cases, some intermediate processing such as logo insertion or picture re-sizing is required or, in many cases, the bit rate has to be dropped. Furthermore, concatenated encoders operating in 4:2:0 mode suffer a loss of vertical chrominance resolution with each compression stage. However, this can be avoided by operating in 4:2:2 Profile mode until the last compression stage. In any case, GOP-aligned encoders produce the best concatenation performance because they minimise the quadrature-phase components of successive noise signals.

14.3   Detection of MPEG frame types in decoded video signals

The concatenation model described in Figure 14.2 has taught us that apart from using serial digital links between decoders and encoders and setting up encoders with similar configurations, we should ideally use GOP-aligned encoders in order to achieve an optimum concatenation performance. This can be accomplished by one of two methods: either we send some encoding meta-data from the preceding decoder to the next encoder or we detect the GOP structure from the video signal itself. As has been explained in Section 14.1, the former is difficult to realise in practice for a number of reasons. Automatic detection of intra-coded frames on the decoded video signal, on the other hand, could solve the problem without the requirement of expensive decoders.

Fortunately, it is possible to detect the GOP structure of a video signal that has been MPEG encoded and decoded again by identifying certain statistical properties of the video signal that are affected by MPEG compression engines in a known manner. For example, bi-directionally predicted pictures are generally more heavily quantised than intra-coded or forward-predicted ones. Further-more, the quantisation matrix used in intra-coded MPEG-2 pictures is usually different to that in P and B pictures. The differences in the coding of the three picture types lead to measurable statistical differences in the uncompressed video signal. Effectively, MPEG compression leaves a 'fingerprint' on the video signal, which can be detected in downstream equipment.

By analysing the input video signal, state-of-the-art encoders can detect whether or not the video signal had been previously compressed. Furthermore, a number of relevant MPEG compression parameters including picture coding types can be derived from the decoded video signal. By avoiding the need for encoding metadata or helper signals, the cost and complexity of systems can be greatly reduced. At the same time, compression noise of the system is mini-mised, thus achieving an overall superior picture quality.

Note that the method of detecting MPEG compression parameters does not rely on any 'hidden' information that a particular upstream encoder may have deliberately placed in the compressed bit stream. Instead, the algorithm detects subtle statistical differences in the video signal itself, which appear after the video signal has undergone MPEG compression and decompression by any compliant encoder and decoder. Therefore, the detection is independent of the type of upstream encoders and decoders. However, overall compression noise is lowest if encoders of the same type are used in the chain. Encoders of the same type tend to generate more highly correlated compression noise because they usually find the same motion vectors and follow the same mode decisions. Therefore, overall compression noise is lower, as outlined in Section 14.2.

14.4   Bit-rate saving

The significant improvements in picture quality shown in Figure 14.3 with GOP-aligned encoders translate into substantial savings in bit rate. However, due to the complexity of concatenated compression systems, a number of assumptions have to be made, in order to derive meaningful figures. For example, assume the concatenation of two MPEG-2 compression systems, both running at 4 Mbit/s, whereby the second encoder is GOP aligned to the first one. If we then misalign the GOP structure that would be most likely to occur in the case of naïve SDI concatenation, we need to increase the bit rate of the second encoder, to achieve the same picture quality again at the output of the second decoder. Table 14.1 shows the required bit rates and bit-rate savings due to GOP alignment based on these assumptions.

It can be seen in Table 14.1 that the results are extremely diverse. This is due to the highly non-linear behaviour of the system. In general, sequences that are non-critical in terms of single MPEG-2 compression also cascade well (Juggler and Popple), whereas critical sequences incur substantial degradation (Mobile and Renata). GOP alignment achieves consistently high picture quality in concatenated systems by improving in particular those sequences that would otherwise incur the greatest degradation.

Table 14.1   *Required bit rate of second encoder with naïve concatenation compared to GOP-aligned concatenation at 4 Mbit/s. The first encoder is running at 4 Mbit/s in both cases*

	Flower Garden	Juggler	Mobile & Calendar	Popple	Renata	Susie	Average
Bit rate (Mbit/s)	5.75	4.13	5.93	4.19	5.94	4.70	5.1
Bit rate saving (%)	43.8	3.3	48.3	4.8	48.5	17.5	27.7

14.5   MPEG-2/MPEG-4 (AVC) concatenation

Most of the analysis outlined in the previous sections is not specific to a particular compression algorithm. For example, the concatenation model can readily be applied to concatenation between different compression algorithms. Furthermore, some compression parameters can be detected in MPEG-2 as well as MPEG-4 (AVC) decoded video signals. Whereas a few years ago, MPEG-2 followed by MPEG-2 compression was the predominant mode of concatenation, today's complex networks often require MPEG-2 encoding and decoding, followed by MPEG-4 (AVC) encoding. Furthermore, MPEG-4 (AVC) to MPEG-4 (AVC) concatenation will become the norm once MPEG-4 (AVC) systems are used for C&D links, as shown in the previous chapter. Transcoding back from MPEG-4 (AVC) into MPEG-2 legacy systems will also be required in some cases.

In this part of the chapter, we will investigate all four concatenated encoding combinations in detail. In order to compare the different concatenation modes, we define concatenation loss as the difference in picture quality between a single and a double encode/decode operation at different bit rates and possibly different algorithms. To derive meaningful results, we operate one of the two concatenated encoders at the same bit rate and algorithm as the reference encoder, while we change the bit rate and algorithm of the other. Figure 14.5 shows a block diagram of the measurement set-up. Encoders 1 and 2 and Decoders 1 and 2 form the concatenation chain, whereas Encoder 3 and Decoder 3 represent the reference chain.

The rest of this chapter is divided into two parts. In the first part, the bit rate of Encoder 1 is held constant while the bit rate of Encoder 2 is varied. This combination is studied at some depth because it represents the more common case, where the bit rate of the upstream encoder is fixed and that of the downstream encoder is set to achieve the required picture quality. A typical example is a constant bit-rate distribution system followed by a variable bit-rate DTH compression in a statistical multiplexing system.

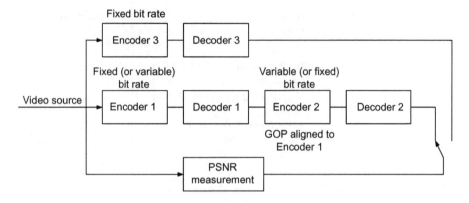

Figure 14.5   Block diagram of measurement set-up

On the other hand, there are many cases in which the bit rate of the downstream encoder is fixed but there is a choice of bit rate and compression algorithm for the distribution encoder. This case is investigated in Section 14.7. Since the behaviour of compression systems is highly non-linear, one combination of encoders, bit rates and algorithms cannot be extrapolated from the other.

14.6   Fixed bit rate into variable bit rate

14.6.1   MPEG-2 to MPEG-2 concatenation

It has been shown in Section 14.4 that MPEG-2 encoders with frame-type detection can be configured to minimise concatenation loss by locking on to the coding structure of the upstream compression engine. This is achieved by identifying video frames that have previously been MPEG-2 intra-coded and by making sure that the GOP structure, GOP length and other coding parameters of the downstream encoder (e.g. field/frame picture coding, etc.) are identical to that of the upstream encoder.

If this is the case and if the two encoders are set up to a similar configuration, the downstream encoder can follow the coding steps of the upstream encoder as closely as possible, thus producing effectively the same compression noise as the upstream encoder. As a result, the noise contributions of Encoders 1 and 2 are highly correlated, resulting in an overall noise power that is not much bigger than that of the larger of the two encoders.

GOP alignment, therefore, makes it possible to reduce the concatenation loss in MPEG-2 to MPEG-2 concatenated systems to pure quantisation noise, i.e. the noise that is due to quantisation mismatch. Certain ratios of quantisation factors result in much greater noise contributions than others. For example, changing the quantisation parameter (QP) from 15 down to 11 produces significantly more transcoding loss than changing it from 15 down to 14, despite the fact that 14 is a coarser quantisation than 11. This is illustrated in Figure 14.6, where the transcoding loss due to quantisation is plotted against one of the QPs with the other held at 15. It can be seen that the total quantisation noise depends on both QPs, whereby certain QP ratios are more favourable than others.

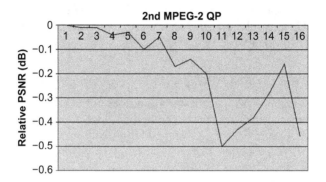

Figure 14.6   MPEG-2 to MPEG-2 concatenation loss for various QPs

In other words, with GOP alignment, MPEG-2 to MPEG-2 concatenation loss depends on relative bit rates and spatial resolutions. For example, if the resolution is dropped from 704 pixels per line to 528 pixels per line, then a corresponding bit-rate reduction of 25 per cent will produce less concatenation loss than other bit rates nearby, even slightly higher ones.

Figure 14.7 shows the concatenation loss of MPEG-2 to MPEG-2 concatenated encoding for two cases: one where the first encoder is set to generate

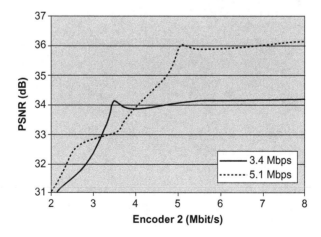

Figure 14.7   MPEG-2 to MPEG-2 concatenation loss

a bit rate of 3.4 Mbit/s and another where it is set to 5.1 Mbit/s. In both cases, all three encoders are set up to full (SDTV) resolution. While Encoder 1 is set to a fixed bit rate, the bit rate of Encoder 2 varies between 2 Mbit/s and 8 Mbit/s. It can be seen that the overall concatenation loss is minimised and picture quality reaches a local maximum when the bit rate of the second encoder is equal to that of Encoder 1. At slightly higher bit rates of Encoder 2, picture quality drops to a local minimum before it reaches the asymptotic limit at 8 Mbit/s. These results and the ones discussed in the following section are based on sports sequences of typical criticality.

14.6.2   MPEG-2 to MPEG-4 (AVC) concatenation

MPEG-2 to MPEG-4 (AVC) concatenation does not exhibit the same bit-rate dependency between the first and the second encoder as MPEG-2 to MPEG-2 concatenation. As can be seen in Figure 14.8, there is no preferred bit rate for the MPEG-4 (AVC) encoder. Picture quality increases monotonically with the bit rate of the downstream (MPEG-4) encoder. This is because there is no correlation between the two noise sources, even if the I frames of the downstream encoder coincide with those of the first encoder and the same GOP structure is used in both encoders. Therefore, the advantage of the higher coding efficiency of MPEG-4 (AVC) encoding manifests itself only if the MPEG-4 encoder runs at a lower bit rate than the MPEG-2 encoder.

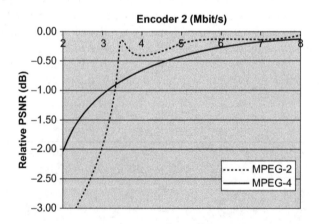

Figure 14.8   MPEG-2 to MPEG-2 and MPEG-2 to MPEG-4 (AVC) concatenation loss

Figure 14.9 shows the concatenation loss of MPEG-2 to MPEG-4 (AVC) concatenation as a function of MPEG-4 quantisation. Again, MPEG-2 quantisation is held at QP = 15. It can be seen that, unlike the MPEG-2 to MPEG-2 case shown in Figure 14.6, there is a smooth transition from the region where coding noise is dominated by MPEG-2 noise (MPEG-4 QP < 19) to the

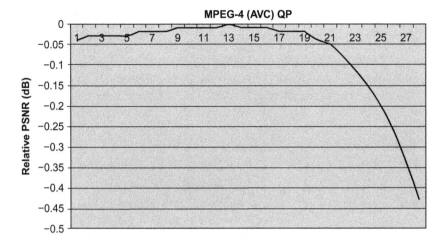

Figure 14.9   MPEG-2 to MPEG-4 (AVC) concatenation loss for various MPEG-4 quantisation factors

region where it is dominated by MPEG-4 noise (MPEG-4 QP > 23). Note that the MPEG-2 QP ranges from 1 to 31, whereas MPEG-4 (AVC) quantisation ranges from 0 to 51.

Figure 14.10 shows how the concatenation loss varies with I frame offset between the two encoders. Both encoders have been set to a GOP length of 12 frames with two B frames between P frames. It can be seen that I frame alignment offers significant benefits in MPEG-2 to MPEG-2 concatenation systems, whereas in MPEG-2 to MPEG-4 (AVC) systems, the effect is far less significant. Bearing in mind that MPEG-4 (AVC) encoders are likely to use different and/or adaptive GOP structures, it is doubtful that any benefit of

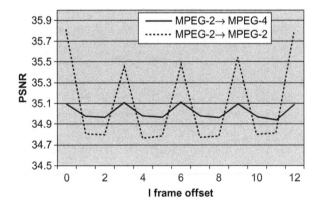

Figure 14.10   MPEG-2 to MPEG-2 and MPEG-2 to MPEG-4 (AVC) concatenation as a function of I frame alignment

I frame or P frame alignment could be achieved in practice. P frame alignment would be accomplished by making sure that the MPEG-4 (AVC) I frame coincides with an MPEG-2 P frame; i.e. MPEG-4 (AVC) I frames would be locked to a local maximum of the MPEG-2 to MPEG-4 curve in Figure 14.10. However, this would restrict the flexibility of the MPEG-4 (AVC) encoder, and it is likely that an unrestricted MPEG-4 (AVC) encoder would achieve a higher picture quality than one that is rigidly locked to an MPEG-2 GOP structure.

14.6.3   MPEG-4 (AVC) to MPEG-4 (AVC) concatenation

Having observed the correlation peak in MPEG-2 to MPEG-2 concatenated systems, one might expect a similar behaviour in MPEG-4 (AVC) to MPEG-4 (AVC) systems. However, the local maximum in picture quality is not as pronounced in MPEG-4 (AVC) concatenated systems. There are several reasons for the reduced correlation between the noise contributions of the two MPEG-4 (AVC) encoders.

- There are significantly more coding modes in MPEG-4 (AVC) than in MPEG-2. Therefore, the chances that the downstream encoder will follow the coding modes of the first encoder are very small indeed.
- The loop filter in MPEG-4 (AVC) acts as a post-processing filter, despite the fact that it is within the prediction loop. This post-prediction filter reduces the prediction (blocking) artefacts, but it also makes it impossible for the downstream encoder not to introduce further prediction noise, even if it makes the same predictions as the first encoder.

Figure 14.11 shows the concatenation loss of MPEG-4 (AVC) to MPEG-4 (AVC) concatenated encoding for two cases with both encoders set up to full

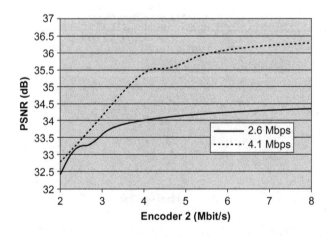

Figure 14.11   MPEG-4 (AVC) to MPEG-4 (AVC) concatenation loss

(SDTV) resolution. Both encoders are set up identically and are I frame aligned. This represents a best case scenario. Encoder 1 is set to a fixed bit rate of 2.6 Mbit/s. The bit rate of Encoder 2 varies from 2 Mbit/s to 8 Mbit/s. Similarly, the second graph in Figure 14.11 shows the case where Encoder 1 is set to a bit rate of 4.1 Mbit/s. Although a slight effect is noticeable at the point where the two bit rates coincide, neither curve shows a significant local maximum, as can be observed in MPEG-2 to MPEG-2 concatenated systems (with GOP alignment).

14.6.4   Concatenation of HDTV

Due to its compression efficiency, particularly in conjunction with statistical multiplexing [6], MPEG-4 (AVC) is already the dominant coding algorithm for HDTV satellite transmissions. Therefore, MPEG-4 (AVC) to MPEG-4 (AVC) concatenation of HDTV signals is becoming increasingly important and it is worth investigating such systems in more detail. In particular, the concatenation performance of 1080 interlaced systems is compared to that of 720 progressive systems.

Figure 14.12 shows the rate-distortion curves of 1080 interlaced and 720 progressive systems for both single encode/decode and concatenated systems. The graphs were generated from a typical test sequence that is available in both formats. It can be seen that to achieve the same picture quality (of 34.5 dB PSNR), a 1080i single encoder requires 13 Mbit/s, whereas the equivalent bit rate of a 720p encoder is 9.5 Mbit/s. In order to achieve the same picture quality through a two-stage MPEG-4 (AVC) concatenated system, the bit rate has to be increased by 24 per cent and 26 per cent for the 720p and 1080i systems, respectively. These figures are based on a test sequence of 'average' criticality.

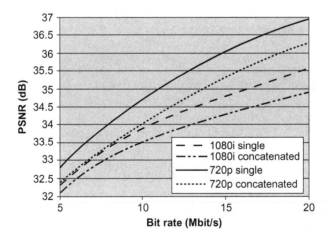

Figure 14.12   MPEG-4 (AVC) to MPEG-4 (AVC) concatenation for different HDTV scanning formats

14.6.5   MPEG-4 (AVC) to MPEG-2 concatenation

The last combination of concatenated compression algorithms to be investigated is from MPEG-4 (AVC) systems back to MPEG-2 legacy systems. Since there are many MPEG-2 systems in operation and MPEG-4 (AVC) is being increasingly used for primary C&D applications, this type of concatenation is becoming increasingly relevant. Unfortunately, this is the worst of the four concatenation possibilities.

Figure 14.13 shows how the picture quality in MPEG-4 (AVC) to MPEG-2 concatenated systems degrades compared to MPEG-4 (AVC) to MPEG-4 (AVC) systems starting with the same initial MPEG-4 (AVC) bit rate of 2.6 Mbit/s.

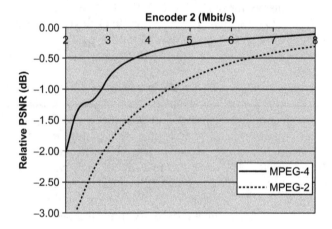

Figure 14.13   Comparison of MPEG-4 (AVC) to MPEG-4 (AVC)
concatenation (solid line) with MPEG-4 (AVC) to MPEG-2
concatenation (dotted line)

Although the concatenation performance of MPEG-4 (AVC) to MPEG-2 is worse than other combinations, this does not mean that MPEG-4 (AVC) should not be used for C&D systems if it is followed by an MPEG-2 system. This is because the coding gain of the MPEG-4 (AVC) encoder exceeds the advantage of an MPEG-2 to MPEG-2 concatenation performance.

14.7   Variable bit rate into fixed bit rate

In Section 14.6, the bit rate of Encoder 1 was held constant while the bit rate of Encoder 2 was varied. This investigated the situation where the compression algorithm and bit rate of the upstream encoder are fixed but the bit rate (and algorithm) of the downstream encoder is chosen to achieve the required quality. In this section, we assume that the bit rate and algorithm of the downstream encoder are fixed but there is a choice of bit rate and algorithm for the upstream encoder.

14.7.1 MPEG-2 downstream encoder

Figure 14.14 shows the rate-distortion curve for both MPEG-2 and MPEG-4 (AVC) compressions in the upstream encoder relative to the distortion of a single MPEG-2 compression at 5.1 Mbit/s. It can be seen that MPEG-2 to MPEG-2 concatenation exhibits the usual local maximum at the point where the two bit rates are similar (with GOP alignment), whereas no such local maximum exists for MPEG-4 (AVC) to MPEG-2 concatenation. In fact, even as the bit rate of the MPEG-4 (AVC) encoder increases, the picture quality barely exceeds that of the MPEG-2 to MPEG-2 case. It is only in exceptional cases that MPEG-2 to MPEG-2 concatenation outperforms MPEG-4 (AVC) to MPEG-2 concatenation, for example when the bit rates and configurations of Encoder 1 and 2 are identical. However, this is not often the case.

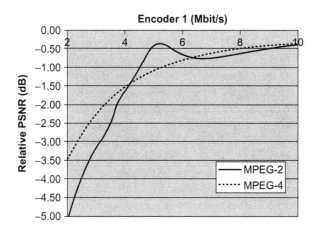

Figure 14.14   MPEG-2 to MPEG-2 and MPEG-4 (AVC) to MPEG-2 concatenation loss

14.7.2 MPEG-4 (AVC) downstream encoder

Figure 14.15 shows the picture quality if the downstream encoder uses MPEG-4 (AVC) compression at 4.1 Mbit/s. This time there is a clear advantage in using MPEG-4 (AVC) compression in Encoder 1 up to a bit rate of about 5 Mbit/s. At that point, the two curves merge and there is very little difference between MPEG-2 and MPEG-4 (AVC) compressions. It is interesting to note that the local maximum for MPEG-4 (AVC) to MPEG-4 (AVC) concatenation is more pronounced in this case than in the case where the bit rate of Encoder 2 is varied (see Figure 14.11).

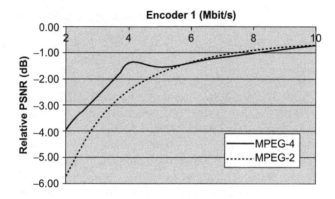

*Figure 14.15   MPEG-2 to MPEG-2 and MPEG-2 to MPEG-4 (AVC)
concatenation loss*

14.8   Concluding remarks

Concatenation of compression algorithms is an increasing problem. One of the
difficulties is that an encoder that is set up for optimum visual quality is not
necessarily set up for optimum concatenation encoding. For example, higher
quantisation on B frames undoubtedly improves the visual quality, but without
a detection of MPEG frame types in downstream encoders, it could lead to a
worse end-to-end performance [7]. Although the bit-rate efficiency of MPEG-2
systems is lower than that of MPEG-4 (AVC) systems, MPEG-2 systems can
be more readily optimised for concatenation performance because there are
fewer encoding parameters to be considered.

   Nevertheless, the higher coding efficiency of MPEG-4 (AVC) systems
improves the performance in all cases where the bit rate of MPEG-4 (AVC) is
lower than that of MPEG-2 systems [8]. This is true regardless of the order of
MPEG-4 (AVC) and MPEG-2 encoding. Since MPEG-4 (AVC) systems are
more likely to be used with flexible and adaptive GOP structures, it is probably
not feasible, nor would it be beneficial, to lock the GOP structure of a downstream MPEG-4 (AVC) encoder to that of an upstream MPEG-2 encoder. It is
probably more important to give the encoders the freedom to achieve the
highest picture quality.

   Apart from the two types of systems considered in this chapter, there are,
of course, many different compression algorithms that a video signal might
undergo before it reaches the end user. Fortunately, most compression algorithms used in studio applications operate at relatively high bit rates and their
impact on end-to-end performance is usually negligible.

   From the examples shown in this chapter, one can conclude that the more
the two compression algorithms differ, the higher is the concatenation loss.
Therefore, to use a different compression algorithm in a given system is of
benefit only if the compression efficiency of the new algorithm is significantly

higher than that of the existing system. This is certainly the case with the MPEG-4 (AVC) algorithm compared to MPEG-2. However, before introducing new algorithms with marginal compression improvements into existing broadcasting chains, the concatenation performance should be carefully evaluated.

14.9 Summary

- MPEG-2 intra-coded frames can be detected in decoded video signals without the use of a helper signal.
- By GOP aligning concatenated encoders, the end-to-end picture quality can be improved.
- While the picture quality degradation of MPEG-2 to MPEG-2 concatenation can be minimised using GOP alignment and identical encoder configurations, this is more difficult to achieve in MPEG-4 to MPEG-4 concatenation due to the large number of coding modes in MPEG-4 (AVC).
- The concatenation performance of 720p is slightly better than that of 1080i.

Exercise 14.1   Concatenation with statistical multiplexing

A broadcast company is planning to upgrade its distribution system to local TV stations. The local TV stations are broadcasting statistically multiplexed SDTV MPEG-2 signals direct to the home. The local MPEG-2 broadcast encoders feature frame-type alignment for optimum concatenation performance. The statistical multiplexing bit rate varies from 1 Mbit/s to 10 Mbit/s. The broadcaster would like to reduce the distribution bit rate without a reduction in picture quality. Should they use state-of-the-art MPEG-2 encoders for the distribution system or MPEG-4 (AVC) encoders?

Exercise 14.2   Concatenation of DTH signals into MPEG-4 (AVC) encoder

An MPEG-4 (AVC) encoder is to be configured in a turn-around statistical multiplex system. The input signals are uncompressed digital video signals that have been MPEG-2 encoded and decoded. The MPEG-4 (AVC) turn-around encoder provides de-blocking and mosquito filters, noise reduction and GOP alignment capability. Which features of the MPEG-4 (AVC) encoder should be enabled or disabled to achieve the best end-to-end picture quality?

Exercise 14.3   Concatenation of DTH signals into MPEG-2 encoder

An MPEG-2 encoder is to be configured in a turn-around system. The input signals are uncompressed digital video signals that have been MPEG-2 or MPEG-4 (AVC) encoded and decoded. The MPEG-2 turn-around encoder provides de-blocking and mosquito filters, noise reduction and GOP alignment capability. Which features of the MPEG-2 encoder should be enabled or disabled to achieve the best end-to-end picture quality?

Exercise 14.4   Concatenation of 4:2:0 signals

Why is it more important to use 4:2:2 compression instead of 4:2:0 in MPEG-2 concatenated systems than in MPEG-4 (AVC) concatenated systems?

References

1. P.J. Brightwell, S.J. Dancer, M.J. Knee, 'Flexible Switching and Editing of MPEG-2 Video Bit Streams', International Broadcasting Convention, Amsterdam, September 1997. pp. 547–52.
2. P.N. Tudor, O.H. Werner, 'Real-Time Transcoding of MPEG-2 Video Bit Streams', International Broadcasting Convention, Amsterdam, September 1997. pp. 296–301.
3. SMPTE Standard 329M – MPEG-2 Video Re-Coding Data Set – Compressed Stream Format, 2000.
4. SMPTE Standard 327M – MPEG-2 Re-Coding Data Set, 2000.
5. A.M. Bock, 'Near Loss-Less MPEG Concatenation Without Helper Signals', International Broadcasting Convention, Amsterdam, September 2001. pp. 222–28.
6. A.M. Bock, 'Statistical Multiplexing of Advanced Video Coding Streams', International Broadcasting Convention, Amsterdam, September 2005.
7. G. Plumb, 'UK HDTV Technical Trials', Conference 'IT to HD 2006', London, November 2006.
8. A.M. Bock, 'MPEG-2/H.264 Concatenation and Transcoding', International Workshop on Digital InfoTainment and Visualization, Singapore, November 2006.

Chapter 15
Bit-stream processing

15.1 Introduction

With the proliferation of transmission systems, e.g. satellite, terrestrial, broadband streaming, transmission to mobile devices, etc., it is inevitable that television programmes are increasingly re-purposed in turn-around systems. We have seen in the previous chapter how the picture quality degradation of concatenated compression systems can be minimised. In many systems the bit-rate step change is so large that the bit-rate reduction can be achieved only by reducing the resolution of the video signal. In this case, decoding followed by down-sampling and re-encoding at a lower resolution is probably the best solution. In some re-multiplexing systems, however, all that needs to be done is to extract a number of channels out of one statistical multiplex group and re-mix them together with some other variable or fixed bit-rate channels into a new statistical multiplex. Assuming that the bit-rate allocation is large enough, decoding and re-encoding can be avoided by using dedicated bit-rate changers.

A second case where decode–encode concatenation can be avoided is switching from one compressed bit stream to another. This can be carried out in the compressed domain if the two bit streams are compatible, i.e. if they have the same picture resolution, the same profile and level and identical audio configurations. Otherwise it is highly likely that there will be a glitch in the decoder. Since the switching process involves video and audio signals, timing references as well as data signals (PSI, SI, EPG, etc.), the process is usually referred to as splicing. Note that although bit-stream processing can avoid decode–encode concatenations, it still requires de-scrambling to get access to header information and (most likely) also re-scrambling.

15.2 Bit-rate changers

There are two types of bit-rate changers: open loop, sometimes referred to as re-quantisation bit-rate changers, and closed loop, also known as motion-compensated bit-rate changers. Open-loop bit-rate changers are much simpler than closed-loop ones, but they have the disadvantage that their bit-rate reduction capability is quite limited. Figure 15.1 shows a block diagram of an

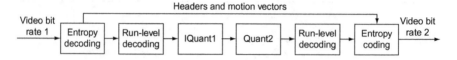

Figure 15.1   Block diagram of an open-loop bit-rate changer

open-loop bit-rate changer. Bit-rate reduction is achieved by increasing the quantisation of the transform coefficients or simply by dropping some of the AC coefficients [1]. As a result of this coefficient manipulation, the decoder ends up with a (slightly) different reconstructed image than the encoder.

As the encoder makes predictions of predictions from the reconstructed images, the divergence between the encoder and the decoder increases with each stage of prediction. The higher the step change in bit rate and the more the prediction generations, i.e. the longer the GOP length, the worse the divergence artefacts. Assuming that the incoming bit stream contains double B frames in between I and P frames, open-loop bit-rate changers can achieve bit-rate reductions of up to about 15 per cent. If, however, the upstream encoder uses adaptive GOP structures with long sequences without B frames, then the bit-rate reduction capability of open-loop bit-rate changers is severely restricted.

The divergence, sometimes also referred to as decoder drift, can be reduced significantly by including a drift compensation loop in the bit-rate changer, as shown in Figure 15.2. This closed-loop bit-rate changer uses inverse DCT transformation and motion compensation to generate new reconstructed images. As a result, the predictions made in the decoder are the same as the predictions made in the bit-rate changer, thus avoiding decoder drift [2].

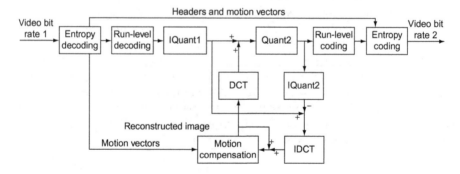

Figure 15.2   Block diagram of a closed-loop bit-rate changer

Closed-loop bit-rate changers can achieve bit-rate reductions of up to about 30 per cent, assuming that the input bit stream is of relatively high quality. Open- as well as closed-loop bit-rate changers reduce the bit rate by re-quantising DCT coefficients. If the input bit stream is already at the lower end of the quality scale (containing coarsely quantised DCT coefficients), further bit-rate reductions might cause severe picture degradation. In extreme cases, a bit-rate reduction might not be possible without a change in picture resolution. Neither open- nor closed-loop bit-rate changers, however, are capable of reducing image sizes.

The drift compensation loop in Figure 15.2 combines the motion compensation loop of a decoder with that of an encoder into a single motion compensation loop. This is possible only if we can re-use the motion vectors of the original bit stream and if the two motion compensation loops are identical. This applies to bit-rate changers, but it does not allow for any other modifications of the bit stream. Even trivial changes such as a change in GOP structure cannot be carried out by a single-loop bit-rate changer as shown in Figure 15.2.

15.3 Transcoding

Changes to bit streams, other than relatively small bit-rate reductions, require a transcoder [3,4]. Transcoders consist basically of a decoder followed by an encoder; i.e. they consist of two fully implemented motion compensation loops: one for the decoder and one for the re-encoder. If the changes in the bit stream are relatively small, the encoder might be able to use a simplified motion estimator by making use of some motion estimation information derived from the decoder [5]; otherwise a full encoder is required.

For larger step changes in bit rates, a reduction in resolution might be required between the decoder and the encoder. Other transcoder applications are profile conversions from 4:2:2 Profile to Main Profile or even a change from MPEG-2 to MPEG-4 (AVC) or vice versa. In terms of concatenation performance, a transcoder is equivalent to a decoder followed by an encoder with GOP alignment, except that the colour conversion from 4:2:0 to 4:2:2 and back again can be avoided.

15.4 Splicing

In the uncompressed video domain, switching from one video signal to another can be carried out seamlessly if the switchover is made in the vertical blanking period. Once a video signal is MPEG encoded, switching or splicing from one compressed video signal to another is not that simple. For a start, there is no vertical blanking period. The only point where a splice into a compressed video bit stream is possible is at the start of an intra-coded picture, and even then one has to observe a number of requirements if a seamless splice is to be achieved.

To enable splicing from one compressed bit stream to another, MPEG [6] and SMPTE [7] have defined a number of requirements for so-called seamless splice points:

- A splice point must start with a *top_field_first*-coded video frame. This applies to the first frame of the bit stream to be spliced into as well as the last frame of the bit stream to be spliced out of.
- If there are B frames immediately after the I frame, no predictions are allowed from the previous GOP.
- The buffer delay at the splice point must be one of a number of pre-defined values between 115 ms and 250 ms, and it must be the same in both bit streams.

Based on these conditions, a number of splicing techniques have been developed [8,9]. Figure 15.3 shows an example of a splice between two bit streams. It can be seen that in the compressed domain, the splice point of the video signal is instigated earlier than that of the audio signal in order to achieve a simultaneous transition in the uncompressed domain. This is necessary because of the longer buffer delay of the video signal in the decoder [10].

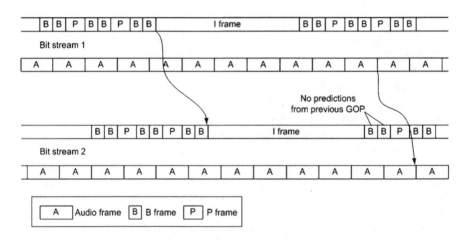

Figure 15.3   Timing diagram of a splice transition in the compressed domain

Even if these requirements are met, there is no guarantee that a compliant decoder will display the splice point seamlessly. The reason is that there are other parameters within elementary and TSs, e.g. video resolution, bit rates, PTS values, etc., that could cause a discontinuity in the bit stream and a disruption in the decode process [11]. If, for example, the video resolution or the audio bit rate is different between the two bit streams, splicing between them will inevitably result in a discontinuity in the bit stream. In these or similar cases, it depends on the behaviour of the decoder how well the splice point is hidden from the viewer. It is only if the splice processing is completely hidden from the decoder that a seamless transition can be guaranteed.

As if the splicing process itself was not complex enough, the control of splicers in real-time broadcast applications is even more challenging. If used for advertisement insertion, for example, real-time encoders, splicers and advertisement servers all need to be controlled simultaneously, in order to synchronise the real-time splice points with the start time and duration of the advertisement interval. This is done using digital programme insertion (DPI) cueing messages as defined in Reference 12.

15.4.1 Splicing and statistical multiplexing

So far we have considered splicing of single programme TSs. In most broadcast systems, however, several channels are combined into one multi-programme TS, often using statistical multiplexing. As has been shown in Chapter 12, statistical multiplexing provides significant bit-rate savings and/or picture quality improvements. However, when it comes to DPI, the fact that the bit rate of the statistically multiplexed programme varies over time makes it difficult to replace it with a (usually) constant bit rate (CBR) local insertion [13].

The problem is illustrated in Figure 15.4. In this example, an advertisement clip encoded at a CBR is to be spliced into a multi-programme TS with four statistically multiplexed channels. The intention is that the advertisement clip is to replace a section of Channel D. It can be seen that the CBR of the advertisement is sometimes higher and sometimes lower than the statistically multiplexed bit rate of Channel D. A possible solution to this problem would be to encode the advertisement clip at the minimum bit rate of Channel D. However, statistical multiplex systems are most efficient if bit rates are allowed to drop to very low on non-critical scenes, almost certainly too low for CBR encoding.

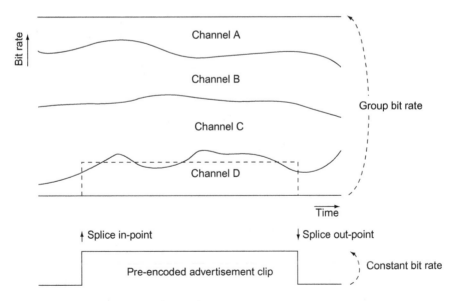

Figure 15.4   Insertion of an advertisement clip encoded at a constant bit rate into a statistical multiplexing group

A better solution is to use bit-rate changers on Channels A, B and C, in order to allocate enough bit-rate capacity for the insertion of the advertisement clip [14].

15.4.2   Insertion of pre-compressed video clips

Splicers often work in conjunction with pre-compressed video sequences, for example advertisement clips. To prepare the video clips requires a content source, an encoder capable of inserting splice points and a control computer with a large storage space. In most cases, compression is carried out using a real-time encoder in order to save time and use conventional video sources such as digital video tapes [15]. Non-real-time software encoders can be used if the content exists in an uncompressed file format and there is enough time available to produce a high-quality video clip.

Figure 15.5 shows a block diagram of an ad insertion system. The ingest controller provides timing and control information to the content source (e.g. digital tape) and the ingest encoder. The encoder extracts the time code from the serial digital interface (SDI) signal and inserts splice points, time codes and I frames at the appropriate points. Since the encoder provides a continuous bit stream with splice points, the controller has to top and tail the captured file before it can be used for ad insertion. Furthermore, the ingest controller adds metadata to the compressed video and audio files before it informs the splicer controller that the ad clip is available. A similar system can be used for video on demand (VOD) ingest.

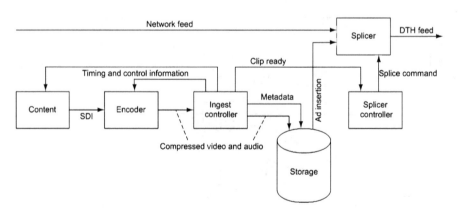

Figure 15.5   Block diagram of an ad insertion system

The network feed itself also contains splice points for ad insertion. The splicer controller selects the ad clip and triggers the splicer to initiate the ad insertion. The splicer then inserts the ad clip at the available splice points of the network feed.

15.5 Concluding remarks

Open-loop bit-rate changers are mostly used in MPEG-2 systems rather than in MPEG-4 (AVC). This is primarily due to the following factors:

- The GOP length in MPEG-2 is usually shorter than in MPEG-4 (AVC), thus limiting decoder drift.
- In order to avoid decoder drift, a significant part of bit-rate reduction has to be carried out on non-referenced B frames.
- MPEG-2 B frames generally contain a sufficient number of DCT coefficients to allow some bit-rate reduction using re-quantisation.
- In contrast, MPEG-4 (AVC) non-referenced B frames contain very few DCT coefficients as a result of advanced inter-prediction modes such as direct mode predictions. Therefore, re-quantisation of non-referenced B frames has little or no effect on bit-rate reduction in MPEG-4 (AVC).

Closed-loop bit-rate changers are suitable for MPEG-2 as well as MPEG-4 (AVC) systems. However, an MPEG-4 (AVC) closed-loop bit-rate changer is considerably more complex than the equivalent MPEG-2 device.

Most transmission systems use conditional access for content protection. Therefore, in addition to the processing blocks shown in Figures 15.1, 15.2 and 15.5, bit-rate changers and splicers need de-scramblers in front of the bit-stream processing and in most cases scramblers afterwards.

15.6 Summary

- Small bit-rate changes can be achieved with bit-rate changers, thus avoiding the need for decoders and encoders.
- Transcoders consist of closely coupled decoders and re-encoders.
- Transcoders are used for large bit-rate or profile changes or conversion from one compression standard to another.
- Splicing between statistically multiplexed video signals requires bit-rate changers.

Exercise 15.1 Bit-stream processing

A telecommunications company is designing a turn-around system for IPTV streaming. It is getting feeds from statistically multiplexed MPEG-4 (AVC) signals with bit rates varying between 250 kbit/s and 5 Mbit/s. The output bit rate is set to a constant 1.5 Mbit/s. Should the company use an open-loop bit-rate changer, a closed-loop bit-rate changer or a decoder followed by an encoder to generate the output bit stream?

Exercise 15.2   Splicing between different bit rates

A splicer is splicing from a 4 Mbit/s CBR MPEG-2 bit stream to a VBR MPEG-2 bit stream with a bit rate ranging from 1 Mbit/s to 4 Mbit/s. All other bit-stream parameters, e.g. horizontal resolution, profile and level, etc., are identical. Can the splicer produce a seamless splice?

References

1. S. Benyaminovich, O. Hadar, E. Kaminsky, 'Optimal Transrating via DCT Coefficients Modification and Dropping', ITRE 3rd International Conference on Information Technology, June 2005. pp. 100–4.
2. T. Shanableh, 'Anatomy of Coded Level Video Transcoding', IET International Conference on Visual Information Engineering, July 2007.
3. D. Lefol, D. Bull, N. Canagarajah, D. Redmill, 'An Efficient Complexity-Scalable Video Transcoder with Mode Refinement', *Signal Processing: Image Communication*, Vol. 22, Issue 4, April 2007. pp. 421–33.
4. E. Barrau, 'A Scalable MPEG-2 Bit-Rate Transcoder with Graceful Degradation', International Conference on Consumer Electronics, 2001. pp. 378–84.
5. P.N. Tudor, O.H. Werner, 'Real-Time Transcoding of MPEG-2 Video Bit Streams', International Broadcasting Convention, Amsterdam, September 1997. pp. 296–301.
6. International Standard ISO/IEC 13818-1 MPEG-2 Systems, 2nd edition 2000.
7. SMPTE 312M Television, Splice Points for MPEG-2 Transport Streams, 2001.
8. W. O'Grady, M. Balakrishnan, H. Radha, 'Real-Time Switching of MPEG-2 Bitstreams', International Broadcasting Convention, Amsterdam, September 1997. pp. 166–70.
9. S.J. Wee, B. Vasudev, 'Splicing MPEG Video Streams in the Compressed Domain', IEEE Workshop on Multimedia Signal Processing, June 1997. pp. 225–30.
10. N. Dallard, A.M. Bock, 'Splicing Solutions in Digital Broadcast Networks', International Broadcasting Convention, Amsterdam, September 1998. pp. 557–61.
11. C.H. Birch, 'MPEG Splicing and Bandwidth Management', International Broadcasting Convention, Amsterdam, September 1997. pp. 541–46.
12. ANSI/SCTE 35 Digital Program Insertion Cueing Message for Cable, 2004.
13. P.J. Brightwell, S.J. Dancer, M.J. Knee, 'Flexible Switching and Editing of MPEG-2 Video Bitstreams', International Broadcasting Convention, Amsterdam, September 1997. pp. 547–52.
14. L. Wang, Y. Liu, B. Zheng, G. Li, 'A Method of Accurate DTV Transport Streams Splicing', International Conference on Consumer Electronics, January 2006. pp. 399–400.
15. G. Berger, R. Goedeken, J. Richardson, 'Motivation and Implementation of a Software H.264 Real-Time CIF Encoder for Mobile TV Broadcast', *IEEE Transactions on Broadcasting*, Vol. 53, Issue 2, June 2007. pp. 584–87.

Chapter 16

Concluding remarks

Over the last 10 years, the compression efficiency of MPEG algorithms has improved significantly, and once all the compression tools of MPEG-4 (AVC) have been fully exploited, it is difficult to see how it could be advanced even further. However, as processing power seems to increase steadily according to Moore's Law, more advanced algorithms can be conceived, and the question is not whether new compression algorithms will be developed, but rather at what point a new standard should be defined.

Apart from MPEG, there are many research programmes and initiatives that could lead to future, more efficient, video compression standards. One of the areas in which compression efficiency could improve on MPEG-4 (AVC) is how to deal with texture with random motion, for example splashing water. This type of content contains little redundancy and is difficult to compress. It has been shown that by synthesising such picture areas rather than trying to compress them, significant coding gains can be achieved [1].

A second potential area for improving video compression is the field of motion compensation. Until now, video compression algorithms have been limited to translational motion compensation. Most instances of natural movement, however, contain elements of more complex motion such as rotation and zoom. With current algorithms, these types of motion have to be broken up into small pieces of translational motion. Once there is enough processing power to search for affine transformations, motion prediction could be significantly improved [2].

These are just two areas where significant improvements in video compression could be achieved. There are, of course, many other areas of video compression research, but it is notoriously difficult to predict which types of algorithms will provide the next breakthrough in video compression.

Apart from video compression itself, automatic picture quality evaluation and monitoring is becoming increasingly important. As the number of digital video channels is increasing, it is becoming more and more difficult to monitor the picture quality of all channels. Whereas state-of-the-art double-ended video quality measurements can readily compete in accuracy and certainly in consistency with the MOS of the ITU-R DSCQS method of subjective video quality evaluation, the accuracy of single-ended video quality measurements remains a challenge.

Apart from the fact that it is difficult to guess what the original video source looked like, high-end broadcast quality encoders are getting better at hiding compression artefacts. Whereas MPEG-1 and MPEG-2 leave distinct 'fingerprints' on the video signal that can be used to estimate picture quality degradations, the in-loop filter of MPEG-4 (AVC) is capable of removing almost all blocking artefacts.

Once texture synthesis is used in video compression systems, picture quality evaluation will become an even greater challenge because a direct comparison with the source will no longer be representative of the picture quality.

References

1. P. Ndjiki-Nya, C. Stuber, T. Wiegand, 'An Improved Video Texture Synthesis Method', International Conference on Visual Information Engineering, London, July 2007.
2. M. Servais, T. Vlachos, T. Davies, 'Bi-directional Affine Motion Compensation Using a Content-Based, Non-Connected Triangular Mesh', 1st European Conference on Visual Media Production, March 2004. pp. 49–58.

Appendix A
Test sequences referred to in this book

There are a number of test sequences referred to in this book. Tables A.1 and A.2 give a brief description of the sequences in terms of content and types of motion.

A.1 SDTV test sequences

Table A.1   Description of SDTV test sequences

Test sequence	Description
Dance	Couple moving in front of landscape background with large plain sky area
Flower Garden	Pan across a flower field with foreground trees and buildings in the background
Football	American football players tumbling onto each other
FTZM	Synthetic images with horizontally and vertically moving captions
Horses	Horses walking in front of still background
Juggler	Juggler with crowd as background and vertically moving captions
Mobile & Calendar	Toy train moving in front of vertically moving calendar and wall paper with highly saturated colours
Popple	Zoom into bird cage and basket of wool with plain blue background
Renata	Renata moving in front of a calendar and wall paper with fine detail
Susie	Close-up of Susie talking on the phone
Table Tennis	Zoom into table tennis player with fine textured background
Tempete	Zoom into natural scene with leaves falling
Tree	Still image of a tree
XCL	Cross colour lady with striped shirt that produces strong cross colour in PAL

A.2   HDTV test sequences

Table A.2   Description of HDTV test sequences

Test sequence	Format	Description
Crowd Run	1080i/720p	Large crowd running with complex motion and high detail
Dance Kiss	1080i/720p	Couple dancing with plain, dark background
Ducks Takeoff	1080i/720p	Close-up of ducks taking off from a lake
Into Castle	1080i/720p	Aerial view of castle and trees zooming into castle
Into Tree	1080i/720p	Aerial view of castle and trees zooming into tree
New Mobile	1080i/720p	Zoom out of calendar with toy train moving in front of plain wall paper
Old Town Cross	1080i/720p	Zoom into an aerial view of an old town
Old Town Pan	1080i/720p	Pan across an aerial view of an old town
Park Joy	1080i/720p	People running in front of trees with detailed texture
Park Run	1080i/720p	Camera following a person running in front of trees
Passing By	1080i/720p	Close-up of people running through a park
Princess Run	1080i/720p	Person running towards camera in front of trees with detailed texture
Seeking	1080i/720p	Close-up of people dancing in park with complex motion
Stockholm Pan	1080i/720p	Pan across an aerial view of Stockholm
Tree Tilt	1080i/720p	Lady jumping off a tree with vertical pan
Umbrella	1080i/720p	Close-up of a person with umbrella in pouring rain
Vintage Car	1080i	Vintage car moving towards camera
Walking Couple	1080i	Couple walking through wood

Appendix B
RGB/YUV conversion

B.1 SDTV conversion from RGB to YUV

$$Y = 0.299R + 0.587G + 0.114B \tag{B.1}$$

$$U = 0.493(B - Y) \tag{B.2}$$

$$V = 0.877(R - Y) \tag{B.3}$$

B.2 SDTV conversion from YUV to RGB

$$B = Y + \frac{U}{0.564} \tag{B.4}$$

$$R = Y + \frac{V}{0.713} \tag{B.5}$$

$$G = \frac{Y - 0.299R - 0.144B}{0.587} \tag{B.6}$$

B.3 HDTV conversion from RGB to YUV

$$Y = 0.2126R + 0.7152G + 0.0722B \tag{B.7}$$

$$U = 0.5389(B - Y) \tag{B.8}$$

$$V = 0.6350(R - Y) \tag{B.9}$$

B.4   HDTV conversion from YUV to RGB

$$B = Y + \frac{U}{0.5389} \tag{B.10}$$

$$R = Y + \frac{V}{0.6350} \tag{B.11}$$

$$G = \frac{Y - 0.2126R - 0.0722B}{0.7152} \tag{B.12}$$

Appendix C
Definition of PSNR

PSNR is calculated by summing up the squared pixel differences between the distorted and the source video signal. It has to be calculated for each component separately.

$$\text{PSNR} = 10 \log_{10} \left(\frac{1}{N} \sum_{i=1}^{N} \frac{255^2}{(V_i - S_i)^2} \right) \text{dB} \tag{C.1}$$

where V is the distorted signal (8 bits) and S is the source signal (8 bits).

Appendix D

Discrete cosine transform (DCT) and inverse DCT

The 8×8 two-dimensional DCT is defined as

$$F(u, v) = \frac{1}{4} C(u) C(v) \sum_{x=0}^{7} \sum_{y=0}^{7} f(x, y) \cos \frac{(2x+1)u\pi}{16} \cos \frac{(2y+1)v\pi}{16} \quad \text{(D.1)}$$

where u, v, x, $y = 0, 1, .. 7$; x, y are spatial coordinates and u, v are frequency coordinates in the transform domain

$$\begin{aligned} C(u), C(v) &= \frac{1}{\sqrt{2}} \quad \text{for } u, v = 0 \\ C(u), C(v) &= 1 \quad \text{elsewhere} \end{aligned} \quad \text{(D.2)}$$

The inverse DCT is defined as

$$f(x, y) = \frac{1}{4} C(u) C(v) \sum_{x=0}^{7} \sum_{y=0}^{7} F(u, v) \cos \frac{(2x+1)u\pi}{16} \cos \frac{(2y+1)v\pi}{16} \quad \text{(D.3)}$$

Appendix E
Introduction to wavelet theory

E.1 Introduction

At first sight, wavelet transformation seems to combine several advantages of sub-band coding and conventional FFT or DCT while being computationally more efficient. The continuous nature of the transform, as opposed to DCT blocks, helps to avoid artefacts, and it appears to be better suited to the spatial de-correlation of texture in images. In the form of quadrature mirror filters (QMFs), a special case of wavelet filters has been known for some time [1]. Wavelet theory generalises the principle of QMFs and provides a broader mathematical basis.

The basic principles of wavelet transformation can be compared to windowed Fourier transformation and sub-band decomposition. Initially, the driving force behind the development of wavelet analysis was the fact that Fourier transformation needs extremely high frequencies for the representation of discontinuities in the time domain. Furthermore, classification of Fourier coefficients for the purpose of speech synthesis, for example, leads to very poor results.

E.2 Relationship to windowed Fourier transformation

Equation (E.1) shows the definition of the continuous wavelet transformation in comparison to the windowed Fourier transformation shown in (E.2).

$$F(a, b) = \frac{1}{\sqrt{a}} \int_{-\infty}^{\infty} \Psi\left(\frac{t-b}{a}\right) f(t)\, dt \qquad (E.1)$$

where $\Psi(x)$ is the analysing wavelet.

$$F(\omega, b) = \int_{-\infty}^{\infty} g(t-b) f(t)\, e^{-j\omega t}\, dt \qquad (E.2)$$

where $g(x)$ is the window function.

Although there are obvious similarities between the two approaches, the differences become apparent when considering the effect of spatial discontinuities on the transform. As far as the windowed Fourier transformation is concerned, the following observations can be made:

- The spatial position of discontinuities, relative to the center of the window, is reflected in the phase of the transform over all frequencies.
- The transform domain consists of a (complex) function in linear frequency.
- Spatial discontinuities lead to very high-energy levels at high frequencies.

As a result of these properties, it can be seen that for an accurate representation of spatial discontinuities, all frequencies of the transform are of similar importance, or, in terms of a discrete transformation, the significance of higher order Fourier coefficients decreases rather slowly.

The multi-scale analysis carried out in wavelet transformation, on the other hand, leads to a logarithmic frequency scale that can, obviously, handle a much wider range of frequencies more easily. Since the frequency scale always decreases by a factor of 2 from one decomposition to the next, the signal is subsampled in each stage of decomposition. In that respect, wavelet transformation is very similar to sub-band decomposition.

E.2.1   Orthonormal wavelets

The idea behind wavelets is that if a function orthogonal to itself, when shifted and compressed by a factor of 2, can be generated, this function can be used as the kernel of a transformation in the same way as $e^{-j\omega t}$ is used in Fourier transformation. The simplest example for such a function is the Haar wavelet that has been known since 1940. The Haar wavelet is $+1$ in the interval $0..1$ and -1 in the interval $1..2$. Unfortunately, the 'square' nature of this simple function has severe disadvantages in many applications. Interestingly enough, this function, together with its $+1$, $+1$ counterpart, can also be regarded as the simplest example of a QMF and even as a 0th-order spline. More about splines is discussed in Section E.2.2.

The first serious attempt at generating wavelets was carried out by Morlet with the aim of investigating seismic signals. Equation (E.3) shows the definition of the Morlet wavelet

$$\Psi(t) = \sqrt[4]{\pi}(e^{-jkt} - e^{-2k^2})e^{-t^2/2} \tag{E.3}$$

where

$$k = \pi\sqrt{\frac{2}{\ln 2}} \tag{E.4}$$

Morlet wavelets are still being used today [2]. Unfortunately, his function was 'not quite' orthonormal, which makes it difficult to find algorithms for perfect reconstruction. Furthermore, the Morlet wavelet is not compactly supported, which means that, in principle at least, infinitely long transformation filters have to be used both for decomposition and for reconstruction.

Orthonormal wavelets can be generated by iteratively applying the dilation equation

$$\phi(t) = \sum_{k=-\infty}^{\infty} c_k \phi(2t - k) \tag{E.5}$$

to find the scaling function $\phi(t)$. The wavelet is then given by

$$\Psi(t) = \sum_{k=-\infty}^{\infty} (-1)^k c_{1-k} \phi(2t - k) \tag{E.6}$$

Orthonormality is achieved by satisfying

$$\sum_{k=-\infty}^{\infty} c_k = 2 \tag{E.7}$$

and

$$\sum_{k=-\infty}^{\infty} c_k c_{k-2m} = \begin{cases} 2 & \text{for } m = 0 \\ 0 & \text{elsewhere} \end{cases} \tag{E.8}$$

The first compactly supported wavelet that is also orthonormal was proposed by Daubechies [3]. Her wavelet is defined by only four coefficients, which makes it computationally very efficient. Since the wavelet is orthonormal, the same filter can be used for decomposition and for perfect reconstruction. The disadvantage of all orthonormal wavelets, however, is that, with the exception of the trivial Haar wavelet, they are all necessarily asymmetric. This can lead to undesirable asymmetric distortion effects when coefficients in the transform domain are quantised or compressed.

There are several methods for generating wavelets, such as

- iteratively applying a dilation equation,
- solving a polynomial equation in the Fourier transform domain followed by inverse transformation and
- finding eigenvalues of an appropriately constructed matrix.

In general, these procedures will not automatically lead to orthonormal wavelets unless some further constraints are imposed.

E.2.2   Wavelets based on splines

Cardinal splines are often used to provide polynomial approximations to an arbitrary function in fixed, integer intervals [4]. The polynomials are generated such that a maximum degree of continuity is maintained across the support points. Continuity in this sense means that not only should the polynomial approximation be equal to the function value at each knot, but also a maximum number of derivatives of the polynomials on either side of the knot should be forced to be equal. For example, a cubic spline has a continuity of 2. This means that as well as the function value itself, the first and second derivatives of the approximating splines on either side of a knot can be made equal.

The advantages of using splines as the basis for wavelet transformation are immediately obvious:

- wavelets based on splines are always compactly supported;
- with increasing order number, splines become progressively smoother, which leads to very smooth wavelets and
- wavelets based on splines are always symmetric.

Unfortunately, wavelets based on splines are never orthonormal (with the exception of the Haar wavelet), so that a different set of coefficients has to be calculated in order to generate wavelets for the reconstruction process. Furthermore, the reconstruction wavelets are not compactly supported; i.e. there remains a reconstruction error, which can be made arbitrarily small by using longer reconstruction filters. The derivation of the reconstruction coefficients involves the solution of a system of linear equations of size $N/2$, where N is the number of coefficients.

It is also possible to use a dual algorithm, whereby the decomposition process is approximated by relatively long filters followed by reconstruction filters that are strictly finite (number of coefficients = order number +2). This would be more appropriate for cases where the signal is encoded once and decoded often.

References

1. V.K. Jain, R.E. Crochiere, 'Quadrature Mirror Filter Design in the Time Domain', *IEEE Transactions on Acoustics, Speech and Signal Processing*, Vol. 32, Issue 2, March 1984. pp. 353–61.
2. C.C. Liu, Z. Qiu, 'A Method Based on Morlet Wavelet for Extracting Vibration Signalenvelope', 5th International Conference on Signal Processing Proceedings, 2000. pp. 337–40.
3. C. Vonesch, T. Blu, M. Unser, 'Generalized Daubechies Wavelets', IEEE International Conference on Acoustics, Speech and Signal Processing, March 2005. pp. IV/593–96.
4. C.K. Chui, *An Introduction to Wavelets*, New York: Academic Press, ISBN 0121745848, 1992.

Appendix F

Comparison between phase correlation and cross-correlation

The phase correlation surface is calculated as follows:

$$PCS(x, y) = FFT^{-1}\left(\frac{FFT(S(x, y))FFT(R(x, y))^*}{|FFT(S(x, y))FFT(R(x, y))^*|} \right) \tag{F.1}$$

where FFT is fast Fourier transform, FFT^{-1} is inverse FFT, * denotes complex conjugate, S is the full search area and R is the full reference area.

By comparison, the cross-correlation surface is

$$CCS(x, y) = \frac{FFT^{-1}[FFT(S(x, y))FFT(r(x, y))^*]}{\text{Norm}(S)} \tag{F.2}$$

where r is the reference macroblock zero augmented to the size of S and Norm(S) is a normalisation factor that makes sure that a perfect match results in a CCS peak of unity.

Appendix G

Polyphase filter design

G.1 Down-sampling by N

Design an FIR low-pass filter with a bandwidth of

$$\frac{f_{si}}{2N} \qquad (G.1)$$

where f_{si} is the input sampling frequency.

The filter can be designed by inverse Fourier transforming the theoretical frequency response of the low-pass filter. Since the theoretical frequency response is a square wave, the inverse Fourier transform generates a $\sin(x)/x$ waveform of infinite length. After transformation into the spatial domain, a window function has to be applied to limit the number of filter coefficients. The resulting filter is applied on every Nth input sample as shown in Figure G.1 ($N = 4$).

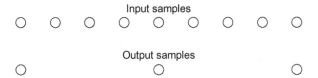

Input samples

Output samples

Figure G.1   Sampling structure of a down-sampling filter with N = 4

G.1.1   Numerical example

$N = 4$, seven-tap low-pass filter, Σcoefficients $= 1\ 024$
    Waveform after FFT^{-1}

..., $-0.045, 0, 0.075, 0.159, 0.225, 0.25, 0.225, 0.159, 0.075, 0, -0.045, ...$

Waveform after applying $\sin(x)/x$ window over seven coefficients

    $0.023, 0.101, 0.203, 0.25, 0.203, 0.101, 0.023$

Coefficients normalised to Σcoefficients $= 1\ 024$

    $25, 115, 230, 284, 230, 115, 25$

G.2   Up-sampling by *M*

Design an FIR low-pass filter with a bandwidth of

$$\frac{f_{so}}{2M} \tag{G.2}$$

where f_{so} is the output sampling frequency $= M \times f_{si}$

The global filter is split into M polyphase filters. As a result, all polyphase filters are all-pass filters. Figure G.2 shows the sampling structure for $M = 4$.

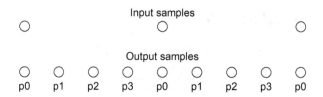

Figure G.2   Sampling structure of an up-sampling filter with **M** *= 4*

G.2.1   *Numerical example*

$M = 4$, 31-tap low-pass filter, Σcoefficients $= 4\ 096$
  Waveform after FFT^{-1}

..., -0.045, 0, 0.075, 0.159, 0.225, 0.25, 0.225, 0.159, 0.075, 0, -0.045, ...

Waveform after applying $\sin(x)/x$ window over 31 coefficients

-0.001, -0.003, 0, 0.008, 0.015, 0.014, 0, -0.023, -0.042, -0.038, 0, 0.071,
0.155, 0.224, 0.25, 0.224, 0.155, 0.071, 0, -0.038, -0.042, -0.023, 0, 0.014,
0.015, 0.008, 0, -0.003, -0.001

Coefficients normalised to Σcoefficients $= 4\ 096$

-4, -12, -15, 0, 32, 61, 56, 0, -93, -170, -155, 0, 289, 633, 914, 1024, 914,
633, 289, 0, -155, -170, -93, 0, 56, 61, 32, 0, -15, -12, -4

Polyphase filters

P0: 0, 0, 0, 1024, 0, 0, 0
P1: -15, 56, -155, 914, 289, -93, 32, -4
P2: -12, 61, -170, 633, 633, -170, 61, -12
P3: -4, 32, -93, 289, 914, -155, 56, -15

G.3   Up-sampling by *M/N*

Design an FIR low-pass filter with the same bandwidth as up-sampling by M
(see (G.2)). Again the global filter is split into M polyphase filters, but the
spatial positions of the output samples relative to input samples is different.
Figure G.3 shows the sampling structure for $M = 4$, $N = 3$.

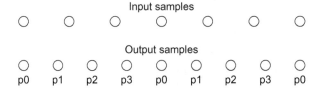

Figure G.3   Sampling structure of an up-sampling filter with M = 4, N = 3

G.3.1   Numerical example

With $M = 4$, $N = 3$, 31-tap low-pass filter and Σcoefficients $= 4\,096$, the numerical example is identical to that of up-sampling by M.

G.4   Down-sampling by *M/N*

Design an FIR low-pass filter with a bandwidth of

$$\frac{f_{sg}}{2N} \tag{G.3}$$

where f_{sg} is the global sampling frequency $= M \times f_{si}$

Again the global filter is split into M polyphase filters. Figure G.4 shows the sampling structure for $M = 3$, $N = 4$.

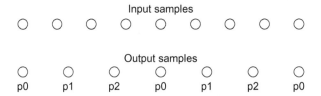

Figure G.4   Sampling structure of a down-sampling filter with M = 3, N = 4

G.4.1   Numerical example

$M = 3$, $N = 4$, 23-tap low-pass filter, Σcoefficients $= 3\,072$
Waveform after applying $\sin(x)/x$ window over 23 coefficients

0.002, 0.006, 0.008, 0, -0.017, -0.034, -0.033, 0, 0.068, 0.152, 0.223, 0.25, 0.223, 0.152, 0.068, 0, -0.033, -0.034, -0.017, 0, 0.008, 0.006, 0.002

Coefficients normalised to Σcoefficients $= 3\,072$

6, 19, 23, 0, -52, -104, -102, 0, 208, 468, 685, 770, 685, 468, 208, 0, -102, -104, -52, 0, 23, 19, 6

Polyphase filters

P0: 23, −104, 208, 770, 208, −104, 23

P1: 19, −52, 0, 685, 468, −102, 0, 6

P2: 6, 0, −102, 468, 685, 0, −52, 19

Appendix H
Expected error propagation time

Assuming that a bit error occurs somewhere in a GOP, the expected error propagation time through the GOP can be calculated as a function of the GOP structure and the relative frame sizes of I, P and B frames. For example, if the error occurs in the I frame, the error propagates through the entire GOP period. However, if the error occurs on a non-referenced B frame, the error is only visible for one frame period. Therefore, the expected error propagation period can be calculated as follows:

$$\text{Error\_time} = \left(\frac{I}{\text{GOP}} N + \sum_{i=1}^{N/M-1} \frac{P}{\text{GOP}} iM + \sum_{i=1}^{N-N/M-1} \frac{B}{\text{GOP}} \right) \text{frame\_period}$$

(H.1)

where N = no. of frames in a GOP, M = distance between reference (I or P) frames, I = I frame size in bits, P = P frame size in bits, B = B frame size in bits, GOP = GOP size in bits.

$$\text{GOP} = I + P\left(\frac{N}{M} - 1 \right) + B\left(N - \frac{N}{M} - 1 \right)$$

(H.2)

If we assume certain frame size ratios P/I and B/I, then (H.1) can be rearranged as follows:

$$\text{Error\_time} = \frac{\text{frame\_period}}{1 + \frac{P}{I}\left(\frac{N}{M} - 1 \right) + \frac{B}{I}\left(N - \frac{N}{M} - 1 \right)}$$
$$\times \left(N + \sum_{i=1}^{N/M-1} \frac{P}{I} iM + \sum_{i=1}^{N-N/M-1} \frac{B}{I} \right)$$

(H.3)

Therefore, the expected error propagation time is a function of the GOP structure (N and M) and frame size ratios.

Appendix I

Derivation of the bit-rate demand model

The bit-rate demand model is derived by defining the cumulative distribution function (CDF). The upper part of the CDF model is given by

$$\text{CDF} = 1 - e^{(B_2 - M_2 \times R)} \quad R > R_{50} \tag{I.1}$$

with

$$M_2 = \frac{\ln(0.05) - \ln(0.5)}{R_{95} - R_{50}} \tag{I.2}$$

and

$$B_2 = \ln(0.5) - M_2 \times R_{50} \tag{I.3}$$

where R_{50} is the median bit-rate demand and R_{95} is the bit-rate demand that is not exceeded for 95 per cent of the time.

The lower (polynomial) part of the CDF is then derived by forcing the following conditions:

$$\text{CDF}(R_1) = 0 \tag{I.4}$$

$$\text{PDF}(R_1) = 0 \tag{I.5}$$

$$\text{CDF}(R_{50}) = 0.5 \tag{I.6}$$

$$\text{PDF}(R_{50}) = -e^{(B_2)} \times M_2 \times e^{(M_2 \times R_{50})} \tag{I.7}$$

In particular, conditions (I.6) and (I.7) ensure that the two parts of the model join up without discontinuities in CDF and PDF. Based on these four conditions, a third-order polynomial can be derived in closed form.

The complete CDF can, therefore, be approximated by

$$\text{CDF} = 0 \quad R < R_1 \tag{I.8}$$

$$\text{CDF} = A_1 + B_1 \times R + C_1 \times R^2 + D_1 \times R^3 \quad R_1 \le R \le R_{50} \tag{I.9}$$

$$\text{CDF} = 1 - e^{(B_2 - M_2 \times R)} \quad R > R_{50} \tag{I.10}$$

and the PDF can be approximated by

$$\text{PDF} = 0 \quad R < R_1 \tag{I.11}$$

$$\text{PDF} = B_1 + 2 \times C_1 \times R + 3 \times D_1 \times R^2 \quad R_1 \le R \le R_{50} \tag{I.12}$$

$$\text{PDF} = -e^{B_2} \times M_2 \times e^{M_2 \times R} \quad R > R_{50} \tag{I.13}$$

The four variables A_1, B_1, C_1 and D_1 can be calculated by solving the following four simultaneous equations:

$$A_1 + B_1 \times R_1 + C_1 \times R_1^2 + D_1 \times R_1^3 = 0 \tag{I.14}$$

$$B_1 + 2 \times C_1 \times R_1 + 3 \times D_1 \times R2_1^2 = 0 \tag{I.15}$$

$$A_1 + B_1 \times R_{50} + C_1 \times R_{50}^2 + D_1 \times R_{50}^3 = 0.5 \tag{I.16}$$

$$B_1 + 2 \times C_1 \times R_{50} + 3 \times D_1 \times R_{50}^2 = \text{PDF}(R_{50}) \tag{I.17}$$

Bibliography

M. Al-Mualla, C.N. Canagarajah, D.R. Bull, *Video Coding for Mobile Communications: Efficiency, Complexity and Resilience*, Academic Press, ISBN 0120530791, 2002.

T. Ang, *Digital Video Handbook*, Dorling Kindersley Publishers Ltd., ISBN 140530636X, 2005.

H. Benoit, *Digital Television: MPEG-1, MPEG-2 and Principles of the DVB System*, Focal Press, 2nd edition, ISBN 0240516958, 2002.

V. Bhaskaran, K. Konstantinides, *Image and Video Compression Standards: Algorithms and Architectures*, Kluwer Academic Publishers, 2nd edition, ISBN 0792399528, 1997.

M. Bosi, R.E. Goldberg, *Introduction to Digital Audio Coding and Standards*, ISBN 1402073577, 2002.

A.C. Bovik, *Handbook of Image and Video Processing*, Academic Press, 2nd edition, ISBN 0121197921, 2005.

C.K. Chui, *An Introduction to Wavelets*, Academic Press, ISBN 0121745848, 1992.

W. Effelsberg, R. Steinmetz, *Video Compression Techniques: From JPEG to Wavelets*, Morgan Kaufmann, ISBN 3920993136, 1999.

C. Fogg, D.J. LeGall, J.L. Mitchell, W.B. Pennebaker, *MPEG Video Compression Standard*, Springer, ISBN 0412087715, 1996.

B. Furht, J. Greenberg, R. Westwater, *Motion Estimation Algorithms for Video Compression*, Kluwer Academic Publishers, ISBN 0792397932, 1996.

M. Ghanbari, *Standard Codecs: Image Compression to Advanced Video Coding*, IET, ISBN 0852967101, 2003.

M. Ghanbari, *Video Coding: An Introduction to Standard Codecs*, IET, ISBN 0852967624, 1999.

L. Hanzo, P. Cherriman, J. Streit, *Video Compression and Communications: From Basics to H.261, H.263, H.264, MPEG4 for DVB and HSDPA-style Adaptive Turbo-transceivers*, Wiley Blackwell, 2nd edition, ISBN 0470518499, 2007.

L. Hanzo, C. Sommerville, J. Woodard, *Voice and Audio Compression for Wireless Communications*, 2nd edition, Wiley Blackwell, ISBN 0470515813, 2007.

R.L. Hartwig, *Basic TV Technology: Digital and Analog (Media Manuals)*, Focal Press, 4th edition, ISBN 0240807170, 2005.

G.H. Hutson, *Colour Television Theory, PAL System Principles and Receiver Circuitry*, McGraw-Hill, ISBN 070942595, 1971.

K. Jack, *Video Demystified, A Handbook for the Digital Engineer*, Newnes, 5th edition, ISBN 0750683953, June 2007.

H. Kalva, J.-B. Lee, *The VC-1 and H.264 Video Compression Standards for Broadband Video Services* (Multimedia Systems and Applications), Springer-Verlag, ISBN 0387710426, 2008.

N. Kehtarnavaz, M. Gamadia, *Real-Time Image and Video Processing: From Research to Reality*, Morgan & Claypool, ISBN 1598290523, 2006.

P.M. Kuhn, *Algorithms, Complexity Analysis and VLSI Architectures for MPEG-4 Motion Estimation*, Kluwer Academic Publishers, ISBN 0792385160, 1999.

I.J. Lin, S.Y. Kung, *Video Object Extraction and Representation: Theory and Applications*, Kluwer Academic Publishers, ISBN 0792379748, 2000.

A.C. Luther, A.F. Inglis, *Video Engineering*, McGraw-Hill Professional, 3rd edition, ISBN 0071350179, 1999.

V.K. Madisetti, D.B. Williams, *The Digital Signal Processing Handbook*, CRC Press, ISBN 0849385725, 1998.

T. Moscal, *Sound Check: Basics of Sound and Sound Systems*, Hal Leonard Corporation, ISBN 079353559X, 1995.

B. Mulgrew, P.M. Grant, J. Thompson, *Digital Signal Processing: Concepts and Applications*, Palgrave Macmillan, ISBN 0333963563, 2002.

N.M. Namazi, *New Algorithms for Variable Time Delay and Nonuniform Image Motion Estimation* (Computer Engineering and Computer Science), Intellect Books, ISBN 0893918474, 1995.

K.N. Ngan, T. Meier, D. Chai, *Advanced Video Coding: Principles and Techniques: The Content-based Approach* (Advances in Image Communication), Elsevier Science, ISBN 044482667X, 1999.

C. Poynton, *Digital Video and HDTV Algorithms and Interfaces*, Morgan Kaufmann, ISBN 1558607927, 2003.

L.R. Rabiner, B. Gold, *Theory and Application of Digital Signal Processing*, Prentice-Hall, ISBN 0139141014, 1975.

R.P. Ramachandran, R. Mammone, *Modern Methods of Speech Processing*, Kluwer Academic Publishers, ISBN 0792396073, 1995.

U. Reimers, *Digital Video Broadcasting: The International Standard for Digital Television*, Springer-Verlag Berlin and Heidelberg GmbH, ISBN 3540609466, 2001.

U. Reimers, *DVB: The Family of International Standards for Digital Video Broadcasting* (Signals and Communication Technology), Springer-Verlag Berlin and Heidelberg GmbH, 2nd edition, ISBN 354043545X, 2002.

I.E.G. Richardson, *H.264 and MPEG-4 Video Compression, Video Coding for Next-generation Multimedia*, Wiley, ISBN 0470848375, 2003.

I.E.G. Richardson, *Video Codec Design: Developing Image and Video Compression Systems*, Wiley, ISBN 0471485535, 2002.

M.J. Riley, I.E.G. Richardson, *Digital Video Communications*, Artech House, ISBN 0890068909, 1997.

Y.Q. Shi, H. Sun, *Image and Video Compression for Multimedia Engineering: Fundamentals, Algorithms and Standards*, CRC Press, Inc., ISBN 0849334918, 1999.

S.J. Solari, *Digital Video and Audio Compression*, McGraw-Hill, ISBN 0070595380, 1995.

M. Sonka, V. Hlavac, R. Boyle, *Image Processing, Analysis, and Machine Vision*, Thomson Learning, ISBN 049508252X, 2007.

V. Steinberg, *Video Standards: Signals, Formats and Interfaces*, Snell & Wilcox, ISBN 1900739070, 1997.

P.D. Symes, *Digital Video Compression* (Digital Video and Audio), McGraw-Hill, ISBN 0071424873, 2003.

D. Taubman, M. Marcellin, *JPEG2000: Image Compression Fundamentals, Standards and Practice*, Kluwer Academic Publishers, ISBN 079237519X, 2001.

A.L. Todorovic, *Television Technology Demystified*, Focal Press, ISBN 0240806840, 2006.

P.N. Topiwala, *Wavelet Image and Video Compression*, Kluwer Academic Publishers, ISBN 0792381823, 1998.

J. Watkinson, *Art of Digital Audio*, Focal Press, 3rd edition, ISBN 0240515870, 2000.

J. Watkinson, *Compression in Video and Audio* (Music Technology), Focal Press, ISBN 0240513940, 1995.

J. Watkinson, *MPEG 2*, Butterworth-Heinemann, ISBN 0240515102, 1999.

J. Watkinson, *Television Fundamentals*, Focal Press, ISBN 0240514114, 1996.

J. Watkinson, *The MPEG Handbook*, Focal Press, 2nd edition, ISBN 024080578X, 2004.

J. Watkinson, F. Rumsey, *Digital Interface Handbook*, Focal Press, 3rd edition, ISBN 0240519094, 2003.

M. Weise, D. Weynand, *How Video Works*, Focal Press, ISBN 0240809335, 2007.

J. Whitaker, *Master Handbook of Video Production*, McGraw-Hill, ISBN 0071382461, 2002.

J. Whitaker, *Standard Handbook of Video and Television Engineering*, McGraw-Hill, 4th edition, ISBN 0071411801, 2003.

G. Zelniker, F.J. Taylor, *Advanced Digital Signal Processing: Theory and Applications*, Marcel Dekker Ltd., ISBN 0824791452, 1993.

Useful websites

http://en.wikipedia.org/wiki/Arithmetic_coding
http://en.wikipedia.org/wiki/Huffman_coding
http://en.wikipedia.org/wiki/Video_compression
http://en.wikipedia.org/wiki/Video_processing
http://www.bbc.co.uk/rd
http://www.chiariglione.org
http://www.ebu.ch/en/technical/trev/trev_home.html
http://www.h265.net
http://www.mpeg.org
http://www.mpegif.org
http://www.wave-report.com/tutorials/VC.htm

Glossary

2T pulse	A sine-squared pulse with a half-amplitude duration equivalent to the video bandwidth
422P	Coding profile which maintains the 4:2:2 chroma resolution
AAC	Advanced audio coding
AC coefficients	All transform coefficients apart from $C_{0,0}$
Affine transformation	Consists of linear transformation, rotation and scaling
Aliasing	Frequency fold-back when a signal is sampled below the Nyquist rate
Artefact	Noticeable image distortion due to compression or other signal processing, e.g. standards conversion
ASIC	Application-specific integrated circuit
Aspect ratio	Horizontal-to-vertical ratio of the size of display devices
ATSC	Advanced Television Systems Committee
AVC	Advanced video coding
AVS	Audio Video Standard, Chinese non-MPEG compression standard
BER	Bit error rate
Black crushing	Non-linear distortion which destroys dark detail
Blu-ray	Next-generation optical disk format for HDTV video
C&D	Contribution and distribution
CABAC	Context adaptive binary arithmetic coding
CAT	Conditional access table
CAVLC	Context adaptive variable length coding
CBR	Constant bit rate
CDF	Cumulative distribution function
Chroma	Abbreviation for chrominance, i.e. colour-difference signals
CIF	Common intermediate format (352×288 pixels)
Conditional Access	Content protection using scrambling algorithms

Criticality	Compression difficulty
CRT	Cathode ray tube
DC	Literally direct current, usually referring to zero frequency
DCT	Discrete cosine transform
Decoder drift	Noise accumulation between intra-coded frames due to mismatch between encoder and decoder inverse transforms
Direct mode prediction	In direct mode predictions, motion vectors are calculated from neighbouring (spatial) or co-located macroblocks (temporal)
DMOS	Difference in mean opinion score
DPI	Digital programme insertion
DSCQS	Double stimulus continuous quality-scale
DSM-CC	Digital storage media command and control
DSNG	Digital satellite news gathering
DTH	Direct-to-home
DTS	Decode time stamp
Dual-prime prediction	A prediction mode for P pictures, which allows field-based predictions using a single motion vector
DVB	Digital video broadcasting
DVB-S	FEC and modulation standard for TV satellite transmissions
DVB-S2	Successor to DVB-S (new FEC and modulation standard for TV satellite transmissions)
DVD	Digital versatile disc
ECM	Entitlement control message
EMM	Entitlement management message
Entropy coding	Lossless data compression exploiting statistical redundancy
EOB	End of block
EPG	Electronic programme guide
Error concealment	Decoder post-processing to make bit errors less visible
Exp-Golomb	Exponential Golomb code for entropy coding
Extremum filter	Non-linear filter to remove impulse noise
Eye height	Measure of integrity of digitally transmitted signal
FEC	Forward error correction
FFT	Fast Fourier transform
FGS	Fine grained scalability
FIR (filter)	Finite impulse response (filter)
FM threshold click	Impulse noise in frequency-modulated signals at low signal-to-noise ratios

FPGA	Field programmable gate array
Gaussian noise	Noise with a probability density function of normal distribution
GOP	Group of pictures, usually preceded by a sequence header or sequence parameter set
HANC	Horizontal ancillary data
HD	High definition
HDTV	High definition television
IBBP coding	Coding structure with two B frames between reference frames
IDR	Instantaneous decoding refresh
IEC	International Electrotechnical Commission
IEEE	Institute of Electrical and Electronics Engineers
IIR (filter)	Infinite impulse response (filter)
Ingest encoder	Compression encoder to record video and audio onto storage media
IP	Internet protocol
IPP coding	Bit stream consisting of I and P frames, i.e. no B frames
IPTV	Internet protocol television
ISO	International Organization for Standardisation
ITU	International Telecommunication Union
JND	Just noticeable difference
JPEG	Joint Photographic Experts Group
JVT	Joint video team between ITU and ISO/IEC (MPEG)
Kell factor	Parameter determining the effective resolution of a discrete display device
LCD	Liquid crystal display flat screen display technology
Letterbox format	Method of displaying a widescreen format on a 4:3 aspect ratio display by inserting black lines at the top and bottom
Luma	Brightness signal
MAC	Video system with multiplexed analogue components
Macroblock	Video block consisting of 16×16 luma pixels
MBAFF	Macro-block adaptive field/frame (coding)
Median filter	Non-linear filter based on selecting the median of neighbouring pixels
Metadata	Auxiliary information about the actual data
MHEG	Multimedia and Hypermedia Experts Group
MHP	Multimedia Home Platform
Middleware	Computer software that connects high-level software applications

Morphology filter	Non-linear filter based on ranking selected pixels in the neighbourhood
MOS	Mean opinion score
Mosquito noise	DCT noise in plain areas adjacent to high-contrast objects
Motion judder	Visual artefact where smooth motion is portrayed as jerky motion
MPEG	ISO/IEC Moving Picture Experts Group
MPEG LA	MPEG Licence Authority
MSE	Mean square error
NAL	Network adaptation layer
NIT	Network information table
Normalisation	Mathematical process of confining to a set range
NTSC	National Television System Committee
Nyquist rate	It is half the sampling frequency
Orthonormal	Two vectors are orthonormal if they are orthogonal and of unit length
PAFF	Picture adaptive field/frame (coding)
PAL	Phase alternating line, 625-line analogue video standard
Pan and scan	Method of adjusting a widescreen format to a 4:3 aspect ratio by cropping off the sides of the widescreen image
PAT	Programme association table
PC	Personal computer
PCR	Programme clock reference
PDF	Probability density function
PES	Packetised elementary stream
PID	Packet identifier
PMT	Programme map table
Polyphase filter	Set of FIR filters with different output phases used for image re-sizing
Posterisation	Noticeable contours in plain areas
PPS	Picture parameter set
PQMF	Polyphase quadrature mirror filters
PQR	Picture quality rating
PRBS	Pseudo-random binary sequence
PSI	Programme-specific information
PSIP	Programme and system information protocol defined by ATSC
PSNR	Peak signal-to-noise ratio
PTS	Presentation time stamp
PVR	Personal video recorder
QAM	Quadrature amplitude modulation

QCIF	Quarter common intermediate format (176 × 144 pixels)
QMF	Quadrature mirror filter
QP	Quantisation parameter
QPSK	Quadrature phase shift keying
QQCIF	Quarter-quarter common intermediate format (88 × 72 pixels)
Quincunx sampling	Sampling pattern in which the samples of successive lines are offset by one pixel
QVGA	Quarter video graphics array (320 × 240 pixels)
RDO	Rate-distortion optimisation
Redundancy	(Information theory) Duplication of information
RF	Radio frequency
RGB	Red green blue
SAD	Sum of absolute differences
Scrambler	A device that manipulates a bit stream to make it unintelligible without the corresponding descrambler
SD	Standard definition
SDRAM	Synchronous dynamic random access memory
SDTV	Standard definition television
SDI	Serial digital interface
SECAM	Séquentiel Couleur Avec Mémoire
SEI	Supplementary enhancement information
SHV	Super hi-vision (7 680 × 4 320, 60 frames/s)
SI	Service information defined by DVB
SIF	Standard input format (352 × 240, 29.97 frames/s; 352 × 288, 25 frames/s)
SMPTE	Society of Motion Picture and Television Engineers
SMS	Short message service
SNR	Signal-to-noise ratio
Spectrum	A range of frequencies
Spline	Special function defined piecewise by polynomials
SPS	Sequence parameter set
SSCQE	Single stimulus continuous quality evaluation
SSIM	Structural similarity picture quality measure
Subcarrier	A modulated signal that is modulated onto another signal
SVC	Scalable video coding
TFT	Thin film transistor flat-screen display technology

Translational motion	Movement of an object without rotation
TS	Transport stream
TV	Television
VANC	Vertical ancillary data
VBI	Vertical blanking information
VBR	Variable bit rate
VBV	Video buffering verifier
VC	Video codec (SMPTE standard)
VCEG	ITU-T Video Coding Expert Group
VCR	Video cassette recorder
VGA	Video graphics array, standard computer resolution (640×480 pixels)
VITS	Vertical interval test signals
VLC	Variable length codes
VOD	Video on demand
VOL	Video object layers
VOP	Video object planes
VQEG	Video Quality Experts Group
VUI	Video usability information
Wavelet	Small wave used to separate high from low frequencies
Web 2.0	Using the World Wide Web for information sharing
White crushing	Non-linear distortion that destroys bright detail
WMV	Windows media video
YCbCr	Digital component format of luma and chrominance signals
YPbPr	Analogue component format of luma and chrominance signals
YUV	Luma and two colour-difference signals

Example answers to exercises

Although many of the answers provided below are clear-cut cases, there are a few exercises for which different opinions are possible. The answers are based on the author's experience. They do not represent the opinion of Tandberg Television, Part of the Ericsson Group. Therefore, the explanations should be regarded as example solutions, rather than definitive answers.

Exercise 5.1   Encoder evaluation

Picture quality should be compared with visual tests or with a picture quality analyser. It cannot be determined from a list of coding tools or from the data sheet.

Exercise 5.2   Bit-rate saving

At low bit rates (or highly critical content), MPEG-2 produces blocking artefacts. A reduction in horizontal resolution not only increases the bit-per-pixel ratio, but the up-sampling filter in the decoder also tends to smooth block edges and thus reduce blocking artefacts. In MPEG-4 (AVC), the in-loop filter softens the picture but avoids blocking artefacts on critical sequences. Therefore, the effect of reduced resolution is less noticeable in MPEG-4 (AVC) than in MPEG-2.

Exercise 5.3   PSNR limit

Since the bit depth of MPEG-2 is limited to 8 bits, the PSNR limit is approximately

$$\text{PSNR}_{\text{max}} = 10 \log\left(\frac{255^2}{0.5^2}\right) \text{dB} = 54.15\,\text{dB}$$

Exercise 6.1   Clip encoding

MPEG-4 (AVC), VC-1 and AVS are, in principle, all suitable for long-GOP QVGA clip encoding. They all use integer transforms, thus avoiding decoder drift on long GOPs. Furthermore, these standards allow motion compensation beyond picture boundaries and ¼ pixel motion compensation. These features improve the picture quality particularly on small images.

Exercise 6.2   Codec comparison

Codecs using block-based transforms can be compared using PSNR measurements under carefully controlled conditions, i.e. the measurement should be carried out at the original picture size and without noise reduction. Wavelet-based algorithms, on the other hand, should not be compared with block-based algorithms using PSNR measurements as pointed out in Chapter 6.

Exercise 7.1   Motion estimation search range

Figure 7.1 shows that 99 per cent of all motion displacements are less than 40 pixels in SDTV. Covering the same motion displacement using single B and double B frames requires a search range on P frames of 80 and 120 pixels, respectively. For HDTV the search range has to be increased by a factor of $1\,920/720 = 2.67$ in order to cover the same visual motion displacement.

Exercise 8.1   De-interlacing

With temporal sub-sampling, all bottom fields are dropped (see Chapter 8). Therefore, repeated top fields in 3:2 pull-down signals are converted to repeated frames. Whereas repeated fields are hardly noticeable, repeated frames produce strong motion judder on translational motion, e.g. camera pan. The problem can be avoided by making sure that all repeated fields are dropped and signalled as repeated fields to the decoder (see Chapter 5).

Exercise 8.2   Vertical down-sampling

Section 8.2.2.1 explains that temporal sub-sampling drops one field of each frame. Figure 2.4 shows that if one field of each frame is dropped in an interlaced signal, there is no detail or motion left, i.e. in both cases the down-converted signal ends up with a stationary signal without vertical detail or temporal motion.

Exercise 8.3   HD to SD conversion

There are two possibilities to convert a 16:9 aspect ratio to a 4:3 format: pan and scan or letterbox. With pan and scan, the 16:9 picture is horizontally cropped to a 4:3 format. The letterbox format keeps the entire picture width but has black bars at the top and bottom. Therefore, the conversion ratios are as follows:

- *Pan and scan horizontal conversion factor*: 1 920 pixels are cropped to 1 440 and then down-sampled by a factor of 2:1 to 720 pixels.
- *Pan and scan vertical conversion factor*: To convert 1 080 lines to 576 lines requires a polyphase conversion filter of 8/15.
- *Letterbox horizontal conversion factor*: To convert 1 920 pixels to 720 pixels requires a polyphase conversion filter of 3/8.

- *Letterbox vertical conversion factor*: To convert 1 080 lines to 432 lines (720 × 432 produces a 16:9 aspect ratio on a 4:3 TV screen) requires a polyphase conversion filter of 2/5.
- *YUV conversion*: Furthermore the Y, U and V signals have to be converted from the HDTV YUV format to the SDTV YUV format (see Appendix B). Substituting (B.10), (B.11) and (B.12) of Appendix B with (1), (2) and (3) results in the following conversion factors:

$$Y_{SD} = Y_{HD} + 0.101581U_{HD} + 0.196077V_{HD}$$

$$U_{SD} = 0.989285U_{HD} - 0.110587V_{HD}$$

$$V_{SD} = -0.072427U_{HD} + 0.983032V_{HD}$$

Exercise 8.4 Encoder optimisation

Noise reduction parameters, pre-processing filter bandwidth and adaptive QP cannot be optimised using PSNR measurements.

- Noise reduction can improve picture quality while degrading PSNR.
- Reducing the filter bandwidth reduces picture detail as well as compression artefacts. The trade-off between picture detail and compression artefact has to be done with visual subjective tests.
- Adaptive QP improves subjective picture quality but tends to degrade PSNR.

Motion estimation search range, GOP length and the number of B frames can be optimised for a particular test sequence using PSNR measurements.

Exercise 9.1 HDTV contribution

The answer depends, of course, on the relative performance of the MPEG-2 and MPEG-4 (AVC) encoders. However, using a state-of-the-art MPEG-4 (AVC) encoder we can assume that MPEG-4 (AVC) achieves a bit-rate saving of at least 40 per cent on average (for a more detailed comparison see Chapter 12). Furthermore, we have seen in Chapter 9 that the bit-rate demand of 1080p50 is about 30 per cent higher than that of 1080i25 to achieve an excellent picture quality. Therefore, the bit rate used for 1080i25 MPEG-2 is more than adequate for 1080p50 MPEG-4 (AVC) encoding.

Exercise 10.1 Compression for mobile devices

In Chapter 10 we have seen that the average bit-rate saving with B frames is about 17 per cent. The number of channels required to fit an extra channel into the same bandwidth can be calculated as follows:

$$n \times bit\_rate = (n+1) \times bit\_rate \times (1 - 0.17) \tag{1}$$

Since *bit_rate* cancels

$$n = (n+1)0.83 \qquad (2)$$

Re-arranging (2) gives

$$n = \frac{0.83}{0.17} = 4.88 \qquad (3)$$

Therefore, five channels without B frames require as much bit rate as six channels with B frames.

Exercise 11.1   Channel change time

To calculate the minimum channel change time, we assume that the decoder finds an I frame immediately after selecting the new channel, i.e. after one frame period. Furthermore, we assume that the new channel is coded at maximum bit rate, and decoder buffer occupancy at the end of the I frame is only 20 per cent of the video buffer size.

$$Min\_channel\_change\_time = 1 \times 0.04 + \frac{0.2 \times 1.2\,\text{Mbit}}{10\,\text{Mbit/s}} = 0.064\,\text{s} = 64\,\text{ms}$$

To calculate the maximum channel change time, we assume that the decoder needs to wait 15 frames before it encounters an I frame. Furthermore, we assume that the new channel is coded at minimum bit rate, and the decoder buffer occupancy at the end of the I frame is 95 per cent of the video buffer size.

$$Max\_channel\_change\_time = 15 \times 0.04 + \frac{0.95 \times 1.2\,\text{Mbit}}{1\,\text{Mbit/s}} = 1.74\,\text{s}$$

The average channel change time is therefore:

$$Ave\_channel\_change\_time = \frac{0.064\,\text{s} + 1.74\,\text{s}}{2} = 0.9\,\text{s}$$

Exercise 11.2   Filter design

Whereas up-sampling polyphase filters can be designed as all-pass filters, down-sampling filters need to have different cut-off frequencies depending on the down-sampling ratio in order to avoid aliasing.

Exercise 12.1   Statistical multiplexing at median bit-rate demand

No. As can be seen in Figure 12.5, the median bit-rate demand as well as the mode bit-rate demand *increases* with the number of channels, whereas the mean bit-rate demand decreases with the number of channels. To gain a benefit from a statistical multiplexing system, it has to be configured to an average bit rate considerably higher than the median bit-rate demand for a given picture quality. Under normal operating conditions, statistical multiplexing provides a bit-rate saving as shown in Figure 12.6. However, if the group bit rate is pushed too low, the benefit of statistical multiplexing diminishes.

Exercise 12.2   Large statistical multiplex system

With 20 channels the PDF of bit-rate demand has already converged to a Gaussian distribution. Adding more channels does not change the shape of the bit-rate PDF. Therefore, the average bit-rate demand for a given picture quality stays the same.

Exercise 13.1   Distribution link

Figure 12.6 shows that even with only three channels, statistical multiplexing can give a small bit-rate saving or an equivalent picture quality improvement. However, reducing the average bit rate with only three channels has the danger of cross-channel distortion. Therefore, it would be advisable to allocate the same average bit rate per channel and use statistical multiplexing to reduce the probability of distortion.

Exercise 14.1   Concatenation with statistical multiplexing

To achieve a significant bit-rate saving without degrading picture quality, an MPEG-4 (AVC)-based distribution system should be selected. MPEG-2 to MPEG-2 concatenation outperforms MPEG-4 (AVC) to MPEG-2 concatenation only if the encoders are set up identically and the bit rates of the concatenated encoders is very similar (see Figure 14.14). However, this cannot be guaranteed in statistical multiplexing systems that vary bit rates continually.

Exercise 14.2   Concatenation of DTH signals into MPEG-4 (AVC) encoder

Most high-end MPEG-4 (AVC) encoders provide hierarchical GOP structures with referenced B frames. These are not compatible with MPEG-2 GOP structures. Therefore, GOP alignment is unlikely to provide a better end-to-end picture quality than optimum MPEG-4 (AVC) GOP structure configurations. To save bit rate in the statistical multiplex system, the MPEG-4 (AVC) encoder should use moderate noise reduction as well as de-blocking and mosquito filters to remove MPEG-2 artefacts.

Exercise 14.3   Concatenation of DTH signals into MPEG-2 encoder

It depends on the picture quality, resolution and compression standard of the incoming MPEG signal.

- If the upstream signal has been coded in MPEG-2 at full resolution and relatively high quality, GOP alignment should be used and noise reduction and de-blocking filters should be disabled.
- If the upstream signal has been MPEG-2 encoded at a low bit rate and/or at a reduced resolution, noise reduction and de-blocking filter can be used to clean up the MPEG-2-encoded signal before re-encoding.
- If the upstream signal has been MPEG-4 (AVC) encoded, neither GOP alignment nor de-blocking filters or noise reduction is required.

Exercise 14.4   Concatenation of 4:2:0 signals

In MPEG-2 Main Profile there is a recommendation for the spatial position of chroma samples as shown in Figure 2.7, but the 4:2:0 to 4:2:2 up-sampling method is not part of the specification. Therefore, different decoders use different 4:2:0 to 4:2:2 up-sampling filters. This can lead to vertical chroma drift in concatenated systems. However, in MPEG-4 (AVC) the vertical position of chroma samples in 4:2:0 mode can be transmitted to the decoder in the video usability information (VUI). Therefore, in MPEG-4 (AVC) 4:2:0 concatenated systems, vertical chroma drift can be avoided.

Exercise 15.1   Bit-stream processing

In Chapter 15 it can be seen that neither an open-loop bit-rate changer nor a closed-loop bit-rate changer can reduce the bit rate by a factor of 4, i.e. from 5 down to 1.25 Mbit/s. This can be achieved only with a decoder followed by an encoder or a transcoder that combines the two modules into one unit. The reason why an encoder can achieve a higher compression ratio than a bit-rate changer is that it can make use of additional coding and pre-processing tools, e.g. a change in GOP structure, de-interlacing, reducing horizontal resolution, etc.

Exercise 15.2   Splicing between different bit rates

Yes, it is possible to produce a seamless splice between two different bit rates. All compliant decoders can decode variable bit-rate streams. Therefore, a bit-rate change from one bit stream to another can be displayed seamlessly, provided that both bit streams are buffer compliant and provide the same buffer delay at the splice points.

Index

Printed in the USA
CPSIA information can be obtained
at www.ICGtesting.com
JSHW011518221024
72172JS00008B/64